KU-166-036

AURORA

MELANIE WINDRIDGE

AURORA

IN SEARCH OF THE NORTHERN LIGHTS

06/09/2016.

To the Highlands Astronomical Society,

Keep looking up!

Melanie Windridge

WILLIAM
COLLINS

William Collins
An imprint of HarperCollins*Publishers*
1 London Bridge Street
London SE1 9GF

WilliamCollinsBooks.com

First published in the United Kingdom by William Collins in 2016

21 20 19 18 17 16
10 9 8 7 6 5 4 3 2 1

Diagrams by Martin Brown based on references from: United States Geological Survey; Robert Fear, University of Southampton; US Air Force and NASA

A catalogue record for this book is available from the British Library.

ISBN 978-0-00-815609-1

Typeset in Lyon Text by Palimpsest Book Production Ltd, Falkirk, Stirlingshire

Printed and bound in Great Britain by Clays Ltd, St Ives plc.

For my family,
with love.

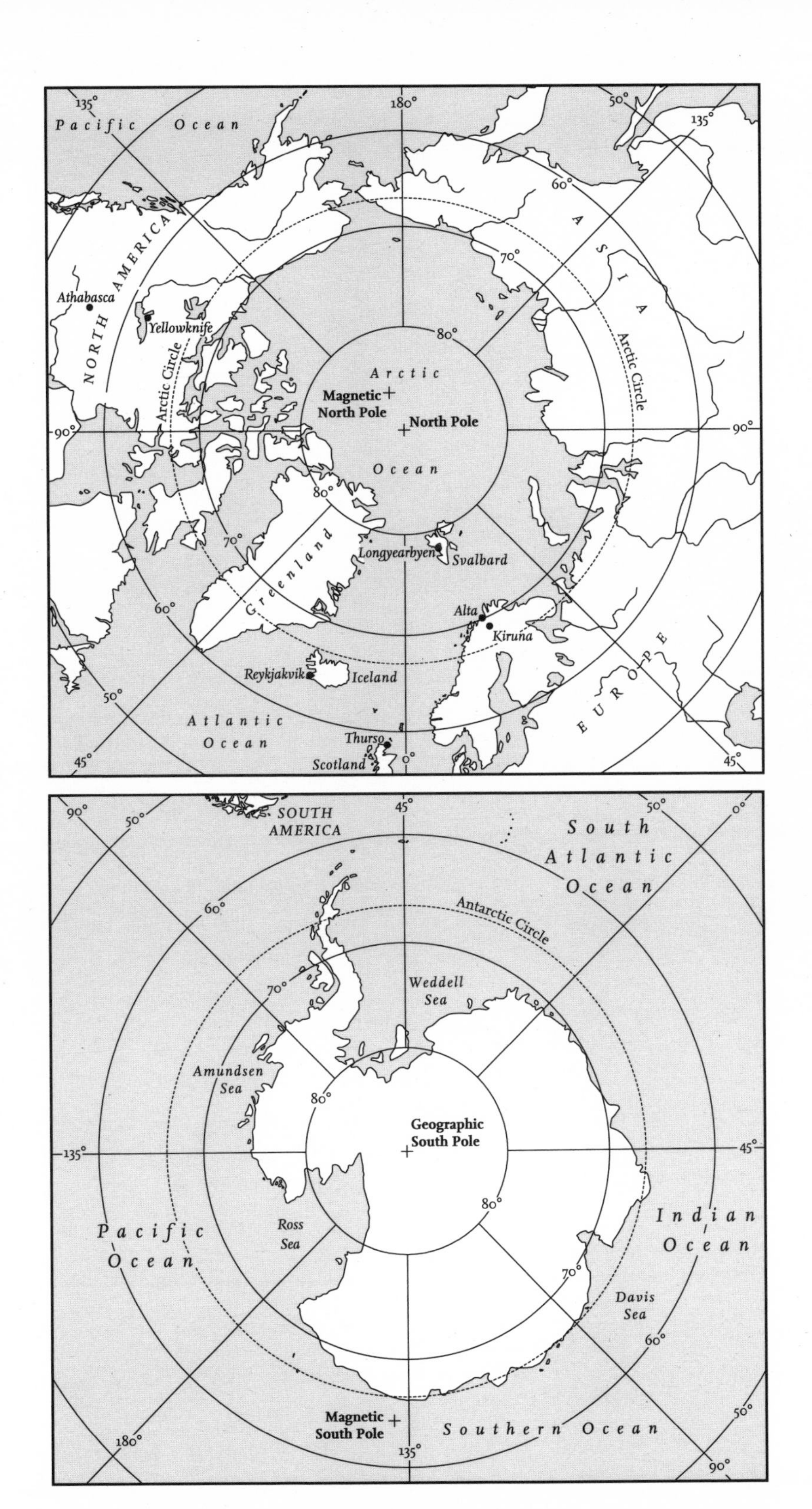

Pacific Ocean
NORTH AMERICA
ASIA
Athabasca
Yellowknife
Arctic Circle
Arctic Ocean
Magnetic North Pole
North Pole
Greenland
Longyearbyen
Svalbard
Alta
Kiruna
Reykjavik
Iceland
Atlantic Ocean
Thurso
Scotland
EUROPE
SOUTH AMERICA
South Atlantic Ocean
Antarctic Circle
Weddell Sea
Amundsen Sea
Geographic South Pole
Ross Sea
Pacific Ocean
Indian Ocean
Davis Sea
Magnetic South Pole
Southern Ocean

CONTENTS

CHAPTER ONE 1
SWEDEN – AT FIRST SIGHT

CHAPTER TWO 23
NORWAY – CULTURE, SPIRITUALITY AND OTHERWORLDLINESS

CHAPTER THREE 57
ICELAND – ROCK, ICE AND FIRE

CHAPTER FOUR 89
CANADA – ELECTRICITY AND MAGNETISM

CHAPTER FIVE 127
CANADA – COLOURFUL COLLISIONS

CHAPTER SIX 161
SCOTLAND – THE DARKER SIDE OF THE LIGHTS

CHAPTER SEVEN 197
SCOTLAND – FORECASTING SPACE WEATHER

CHAPTER EIGHT 251
SVALBARD – SEEING THE LIGHT

FURTHER READING 295

INDEX 297

ACKNOWLEDGEMENTS 305

CHAPTER ONE

SWEDEN – AT FIRST SIGHT

THE NIGHT WAS a velvet canvas scattered with stars. Pillars of colour streaked down from above; green fringed with red and yellow disappearing up into the heavens. The lower edge twisted into a bright band against the darkness. Slowly the colours rippled and broke and reappeared elsewhere, until, suddenly, the whole sky exploded into colour like paper bursting into flames. Here were the northern lights in spectacular form.

It seemed this dance of such dynamic energy was stretched above the whole Earth; the individual dancers moving to their own tunes but somehow in harmony. Twirling and writhing, rippling and pulsing, the shapes rose and fell, sometimes exploding across the sky in a delicate shower of pinks and purples within the dominating green. It was as if a thousand people whirled with green silk sashes and scarves, all different shapes and sizes, swooshing and billowing independently. It was as dramatic as a thunderstorm, yet calm. Gentle, yet astonishing.

The northern night sky is not always this active. There are times of glorious intensity and times of pale tranquillity. There are nights when there is colour and motion; others when there is only a vague streak across the sky and any movement is barely discernible. There are places on Earth where this wild sight is commonplace, the calm colours barely remarked upon; yet for the majority of the world's population the lights are a novelty, a rarity or an irrelevance, all depending on one's location relative to the polar regions.

The northern lights is just one of a pair of twin phenomena. The aurora, or polar lights as it may be generically called, occurs both in the northern and southern hemispheres in ringed regions around the Earth's poles – or, more specifically, around the Earth's magnetic poles. The regions do sometimes expand, giving mid-latitude residents a glimpse of the magic, but, for the most part, to see the lights we need to travel north. So we head to the auroral zone, this narrow band of latitude where the northern lights are a regularity – to northern Scandinavia, Siberia, Canada or Alaska. To seek the southern lights means a trip to Antarctica, unless the night is one of elevated auroral activity when we might see a colourful haze in the south of New Zealand, or in rare, extreme circumstances when the aurora moves equatorwards.

The aurora has a reputation for being elusive, which adds to its mystery and attraction. It is a natural process, completely out of our control. To see it, you must be in the right place at the right time. The conditions must be favourable, too; the skies should be dark and clear. Clouds mask the view, and stray light – even moonlight – can wash out the delicate colours of the aurora. To see it in full splendour is a gift: a coveted, capricious gift.

The need for darkness makes the aurora a winter-time treat. In the summer months in the high-latitude locations where the aurora is created, the top of our planet is tilted directly towards the Sun, always in range of its light. At the Arctic Circle the Sun never sets between mid-June and the beginning of July, and the Earth's poles see this midnight sun for half the year. Even further south of the Arctic Circle – towards the south of Scandinavia, the northern reaches of Scotland, or central Canada – the sky never darkens sufficiently in summer to see something as diaphanous as the aurora.

This winter-time nature adds to the aurora's mystique. It is at one with the cold, snow and ice. It becomes part of the landscape and so, for me, this landscape is very much part of the aurora. So, too, are the people who live there; part of a human legacy stretching back for millennia. The sky connects us to our ancestors.

I am a physicist and my fascination with the northern lights grew gradually to a point where I didn't just want to see it, I wanted to know it. I knew the basic science behind the aurora – that it is an event caused by charged particles that are channelled down magnetic field lines and interact with our atmosphere. But there had to be so much more to it than that. What caused the differences in colour? Why was the aurora predominantly green but then at other times showed flashes of mauves and violets? What caused the various shapes and patterns in the aurora? Why did we sometimes see pillars of light stretching skyward, sometimes twists and ribbons? Why is it sometimes calm and sometimes wild? What causes the sudden eruption in movement and colour? Why does it sometimes move further south? It seemed I still had so many questions.

More than that, I had read accounts of old, heroic polar explorers' experiences of the aurora and noticed the overwhelming feelings of awe and spirituality. I saw how these feelings mirrored much of the ancient folklore of the northern lights. It is only in the last century that science has come up with a plausible explanation for the aurora and I was curious about the effect of such knowledge. Do modern polar explorers feel the same as those first few men, I wondered, or does science overcome the purity of the experience?

I had seen the pictures, watched the videos, heard the stories. I felt the draw of the Arctic – the light and space, the open wilderness, the long winter darkness and the awe-inspiring aurora borealis. I wanted to stand captivated under a wide sky, watching the heavens move in a graceful, barely choreographed dance. I wanted to feel it.

I imagined looking up and seeing the starburst of an auroral corona above me. I imagined colour and light and form. I imagined the whole length, breadth and depth of the sky dominated by arcing light. I imagined a silken landscape of soft snow, where the snowflakes lap over your skis and pile high on the fir tree branches, weighing them down. I imagined dragging sleeping bags out from the tent so we could lie outside on the snow, warm and marvelling at the dancing sky above us, just as one would the stars in the desert.

I never imagined the cold, or the pain.

* * *

THERE WAS A thin covering of snow on the ground as my plane waited at Stockholm airport. Grass poked through it in tufts and there were small, scraped piles of greying slush at the edge of the runway. From the aeroplane window Stockholm appeared in monochrome: the terminal building white with its black details and greying edges; the sky white; the snow on the dark tarmac the same hue. Men in their fluorescent yellow jackets stood out brightly, as if an old film were slowly merging into Technicolor. As we took to the skies I studied the trees and the small snowy fields below, divided up by narrow roads. An irregular patchwork of whites, greys and pale browns, their textures beneath the snow stood out as much as their colour. Almost immediately we were into the clouds and once again everything was white.

I was flying to Kiruna, in Arctic Sweden, to see the northern lights – the aurora borealis – for the first time. For years I had dreamed of seeing it, and finally there I was, en route to the Arctic, a place that held for me as much fascination as the aurora itself. I am a plasma physicist with a visceral attraction to mountains, ice and snow, whilst my academic research had focused on nuclear fusion as a future energy source. In my research days I was motivated by the youthful optimism that fusion would change the energy domain, and provide energy security and a way to temper the threat of climate change that could devastate the precious landscapes that I loved and the communities within them. I was – and still am – attracted by the practicality of this research, but in the aurora my passions converged. The plasma physics danced above the boundless snowy land that held me in thrall.

My fascination with the aurora wasn't new; it had, in fact, dawned on me rather slowly. When I was an undergraduate student I spent one summer holiday working at the Rutherford Appleton Laboratory in Oxfordshire. There I worked on a project that looked at the connection between the Sun and the Earth, and it was there that I learned about the aurora – a captivating light show in the upper atmosphere of the polar regions and a product of the intricate interplay between these two celestial bodies. The name 'aurora' comes from the Latin for 'dawn', and 'borealis' from '*boreas*', meaning the north wind. The term is believed to have been first used as a metaphor by the Italian scientist Galileo Galilei when he referred to the appearance of a bright aurora in the northern parts of the sky as a 'northern dawn'. Similarly, an aurora seen in the southern hemisphere is referred to as 'aurora australis' from the Latin '*auster*', for south wind.

Whilst at the Rutherford Appleton Laboratory I was also lucky enough to watch the live launch of a spacecraft – not live in Kazakhstan, obviously, but in real-time from the lab's main lecture theatre. The rocket took two satellites from the Cluster II mission up into orbit that day; two others had been launched a month previously. As I sat and watched the launch it felt extraordinary; I realised that at that very minute that rocket was leaving the Earth and going into space. The realisation of the otherworldliness of it all thrilled me. Scientists were on tenterhooks because this was the mission's second attempt – the first Cluster spacecraft had been lost in launch in 1996, four years previously. This time the launches were a success and the four identical satellites began their long dance around Earth. They are still flying in formation now, collecting three-dimensional data on near-Earth space and the conditions that cause the aurora. The experiences of that summer stayed with me, imprinting on my mind a latent fascination with our connection with the Sun.

More than ten years later, in northern Sweden on an Arctic Science course, I was about to get my first glimpse of the aurora. This was just the start of my journey, and when I got on that plane to Kiruna I didn't realise quite how far I would go. It was a journey that would take me back to the north several times, each visit piecing together a story of the aurora borealis that at its core was the science behind the phenomenon, but which was woven with landscape, people and history. This is that story.

That early trip to Kiruna was also to be my first experience of the Arctic. I was excited, of course, but I didn't really know what to expect. I was not going out into the wilderness, so it would be a gentle introduction, though our course leaders warned us that temperatures could drop to -25°C and that we should dress appropriately. I didn't know exactly what they meant by that, never having had to dress 'appropriately' for such temperatures before. I had packed mostly ski wear and extra jacket layers, and I was intrigued to know what the Scandinavians wear indoors, when they take off all these layers of outerwear. Did they all walk around in their thermals?

It was a short flight and we were cruising at just 3000 metres (10,000 feet) above sea level. Now, high above the clouds, the sky was a beautiful clear blue. Beneath us the clouds created a thick, woolly covering, stretching on for what seemed like forever. I wondered how much higher I would have to be before I would see the curvature of the Earth, before I would reach the edge of space itself.

However, the divide between Earth and space is blurry rather than sharp. People talk about flying to the edge of space in a fighter jet, or sending a camera up on a weather balloon, but that is in fact only to around 20 kilometres (12 miles) above the Earth's surface. For aerospace purposes, the edge of space is defined as 100 kilometres (60 miles) above sea level – known as the Kármán line, after the Hungarian-American physicist Theodore von Kármán. That is where the atmosphere becomes so thin that aerodynamic forces can no longer provide lift. But that is not even the edge of our atmosphere.

Go quite a lot higher, up to the International Space Station, and even there the Earth looks large and close; only a small part of her surface is visible but the curvature is very apparent. The International Space Station, or ISS, an artificial satellite which has been continually inhabited by astronauts since November 2000, orbits at an altitude of 400 kilometres (250 miles). Even that is still within the Earth's atmosphere, albeit a rapidly thinning one. Our planet's atmosphere gradually peters out into space from about 500 kilometres (300 miles) above the Earth's surface. Astronauts on the ISS have taken photographs and video of the aurora from above – a view from what we call 'near-Earth space'.

As we flew on to Kiruna I reflected on Earth and space – the connections, interactions and varying boundaries between the two systems. Both come into the story of the northern lights, and our humble atmosphere plays an important part. The atmosphere is the screen upon which the drama plays out.

The atmosphere has five main layers, defined according to air temperature, and almost every atmospheric phenomenon with which we are familiar is restricted to the lowest of these, the troposphere. Three-quarters of the mass of the atmosphere is contained in this thin layer, which extends up to an average of 12 kilometres (7.5 miles). The air becomes increasingly rarefied the higher we climb. Because the atmospheric layers are demarcated by temperature, the height of the troposphere is slightly variable across the globe. It is thickest towards the tropics, where warm temperatures cause expansion of the air, and thinner at the poles.

The troposphere is named after change, or turning. The air that is warmed at the surface of the Earth rises and mixes with the overlying air, creating convection currents that generate our weather. Fluffy, white, picture-perfect cumulus clouds are found at around an altitude of only 3 kilometres (1.8 miles) or lower. Towering, thunderous cumulonimbus can stretch from below this height up to almost 15 kilometres (9.3 miles) in giant, peaky columns. Wispy cirrus are blown around at high speed at altitudes between 6 and 12 kilometres (3.5 and 7 miles), about the same altitude at which a commercial jet airliner cruises. The average temperature of the air in the troposphere decreases steadily with altitude to around -60°C. At around 12 kilometres (7.5 miles) up the temperature stabilises and turns, and we enter the stratosphere.

Here, the temperature begins to increase again due to the presence of ultraviolet-absorbing ozone. This ozone layer in the stratosphere restricts the mixing of air, which is why most of the weather we know is confined to the troposphere. The ultraviolet radiation that the ozone layer absorbs increases the temperature back up to around freezing again – a warm zero degrees!

As I squinted out of the window into the bright blue sky and looked down on the clouds, I mused over our sense of scale. Above the clouds it felt as though we were flying so high and yet we were actually so low, in relative terms. I had climbed mountains higher than this before – up to 6000 metres (20,000 feet) – and I planned to go higher. Commercial tandem skydivers would jump from this aeroplane's altitude, and the world record for a parachute jump was thirteen times higher. In October 2012 the Austrian skydiver Felix Baumgartner had jumped from the stratosphere (from 39 kilometres (24 miles) above sea level), having been taken up in a capsule suspended below a large helium balloon. He wore a pressure suit that allowed him to survive in the rarefied air and his descent lasted around ten minutes, over four of those minutes spent in freefall. Yet in the stratosphere we still haven't reached the height of the aurora.

At about 50 kilometres (31 miles) the temperature starts to drop again. From here to 80 kilometres (50 miles) up is called the mesosphere. It is the layer of shooting stars, where meteors coming in towards Earth from space burn up due to the pressure and heat created from collisions with air particles. Yet the air is already remarkably thin, and the top of the mesosphere is the coldest part of the Earth's atmosphere.

Beyond about 80 kilometres (50 miles) altitude is the thermosphere, a layer that is several times thicker than all the others put together. It transitions into the exosphere somewhere around 500 kilometres (300 miles) altitude, but this level can vary wildly depending on the activity of the Sun. If the thermosphere gets hotter it can puff up to as much as 1000 kilometres (600 miles) from Earth. Beyond the thermosphere, the exosphere is where the Earth's atmosphere finally blends out into space over thousands of kilometres. It is a murky transition. The thermosphere, seen as the fourth layer of our atmosphere from one perspective, may also be seen as space from another perspective – it is in this layer of the atmosphere that the International Space Station orbits.

Up here, the number of air particles is rapidly dwindling. Even at the mesosphere–thermosphere transition at 80 kilometres (50 miles) the density is a hundred-thousandth of the density at sea level. A gas particle here can travel about one metre on average before hitting another particle. At sea level, that same gas particle would travel less than the width of a human hair before colliding with another. Up at 200 kilometres (120 miles) altitude, in the thermosphere, the air is so thin that a particle could travel 4 to 5 kilometres (2.5 to 3 miles) without hitting another one. These extreme conditions, the likes of which have never been achieved on Earth, allow ordinary atoms and molecules to behave in unusual ways. This rarefied zone is the three-dimensional canvas for the aurora.

In the thin air of the thermosphere, particles are ionised by solar radiation. Incoming photons of light from the Sun dislodge electrons from their atoms so they move around freely, creating a charged gas of negatively charged electrons and positively charged ions, the remnant atoms. During daytime, some ionisation occurs towards the top of the mesosphere too, and it also occurs beyond the thermosphere, in the exosphere. This multi-layered shell of ionised gas, reaching from about 85 kilometres to 600 kilometres (53 to 370 miles) above the Earth's surface, is known as the ionosphere, and it is here that the aurora is created, predominantly between 100 and 250 kilometres (60 and 155 miles). Pictures of

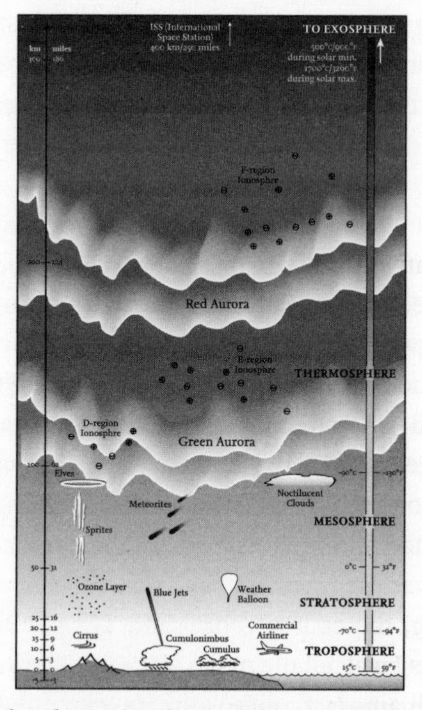

The Earth's atmosphere has various layers, defined according to temperature. Most of the weather we experience is confined to the troposphere, the first 20 km (12 miles), but the aurora occurs much higher up – predominantly around 100 km (60 miles) high but stretching up hundreds of kilometres more.

aurorae taken from the International Space Station show the clear curtains of light stretched out in a bright green that fades up into red, and then fades out altogether.

Seen from this vantage point in near-Earth space, these curtains of colour clothe our planet. From some reports of the aurora, the lights may even seem close enough to touch, yet in reality they are far, far away from us. The aurora may bridge the boundary between Earth and space, straddling the human-imposed divide of the Kármán line.

I was still daydreaming about whether the aurora was an Earthly or space phenomenon when we started to drop closer to the woolly mass of clouds beneath us. The captain came onto the loudspeaker to declare landing in twenty minutes. It was time to stow away the tray tables and turn off all electronic devices. My first glimpse of the Arctic awaited me just on the other side of those clouds.

* * *

KIRUNA IS A relatively new town in European terms, and it is also the least-populated area in Europe. It sprang up in around 1900, when mining began in these parts. The resource here is iron ore, and the town is noted mainly for the mine, its rocket range and space studies, and – more recently – the Icehotel, a building made entirely of ice. The Icehotel melts every spring so every year a new one is built, the hotel reborn in the season of renewal. It is a beautiful mix of science and art, built by engineers and sculpted by artists from all over the world.

Carol Norberg, the course coordinator, was at the airport to meet my flight. As I walked down the steps of the plane, breathing in the chilly, fresh air, my first thought was that it wasn't quite as cold as I had feared. It was actually -6°C, not too cold, just yet. The clouds were keeping us warm. I stepped down into snow, hard-packed, white and creaky. We walked the short distance from the plane to the terminal building – a bold red box with grey-rimmed glass doors that stood out against the white roofs. The terminal building was a single room, about the size of a small school hall, with a snaking black conveyor belt at one side and people milling all around. Almost immediately I saw Carol holding a sign.

Carol herself had lived in Kiruna for several years until recently, when she had moved to Stockholm. As we drove into town on the perfectly white flattened-snow roads she told me about the area. It has an unusual and uncertain future; there are plans to move the town of Kiruna to a new site a few kilometres away, constructing a new town centre out of the way of the encroaching mine. This is because Kiruna is threatened by the very magnetite mine to which the town owes its existence. It is a huge, complex network, and as a result there are as many roads underground in Kiruna as there are ordinary roads in Stockholm. The deposits of iron ore are found in a large slab that cuts down diagonally beneath the town. Mining this ore would cause the ground above, and the town itself, to collapse, so Kiruna must go.

The railway tracks were the first to move, and a construction and relocation process that could take decades is now under way. I asked Carol whether the town's residents were resistant to the upheaval that moving would cause. She explained that many people who live in Kiruna work in the mine. The work pays well and if the town doesn't move then people will lose their jobs. 'Most people are happy to move,' Carol told me. 'They will build a new Kiruna.'

She explained that some of the oldest buildings, like the church, would be moved to the new town so they wouldn't lose their history. We passed a large brown rectangular building, quite unremarkable except for its industrial-style clock tower made from seemingly random bits of metal, the gold clock hands standing out against the dark rust of the tower. 'This is the town hall,' said Carol, 'and it will be one of the first buildings to be evacuated as it is so close to the mine.'

The town will not be moved all at once. The road and rail infrastructure must be set up first, then the municipal facilities. After that, there is a planned itinerary for the movement of the rest of the town, with important historical buildings like Kiruna Church being moved first, followed by groups of residents year by year. Moving a city, albeit a small one like Kiruna, is no small feat. I wonder how it will affect the residents who must move, as they watch their town fragment – its heart literally ripped out – whilst they are forced to stay, perhaps for years, in the shell of the town. Kiruna already faces a shrinking population; the young people move away for better education or more dynamic urban environments. It remains to be seen whether the movement of Kiruna will reinvigorate the town or choke it.

I pondered this as we continued past the brightly coloured

wooden buildings heaped under so much snow that they were only half visible. Even their steeply pitched roofs held the snow. They were, Carol said, typically Swedish, called 'ink-pot houses' because of their shape. Kiruna's more modern buildings are boxy and industrial. Some even have balconies that are designed to look like lifts from the mine. In fact, almost everything in Kiruna is inspired either by the mine, or by space. It is a town of science and industry. The eyes and thoughts of the residents are focused either down below ground or up into the skies.

The following day, some newly met housemates and I made our way through the dim, unfamiliar streets to the bus stop. The sun was rising, flame-like, on the horizon as the bus pulled up, and about fifty students in bright coats and backpacks made a scrum around the door. The bus delivered us to the Swedish Institute of Space Physics, located slightly out of town and surrounded by fir trees heavy with snow.

The Swedish Institute of Space Physics was originally set up as the Kiruna Geophysical Observatory in 1957, with its main activity to be auroral observations. Kiruna was chosen because of its location beneath the auroral oval – the region where the northern lights are seen – and because it already had good infrastructure on account of the working mine. Work began on auroral observation, mostly using sounding rockets launched from the nearby rocket range, but gradually expanded to exploring other processes around the Earth and into space. It turned out that Kiruna was also well placed for atmospheric studies because it sits beneath the polar vortex – a huge low-pressure region of rotating air that forms over the Arctic (and similarly in the Antarctic). The observatory became a state-owned research institute in 1987, changing its name to the Swedish Institute of Space Physics.

There are various other space-based activities around Kiruna, some of which work in collaboration with the Institute. A radar antenna called EISCAT, the European Incoherent Scatter radar, is used to study disturbances in the upper atmosphere. The

EISCAT Scientific Association is headquartered in Kiruna. There is a European Space Agency satellite station that supports ESA satellites in orbit and the rocket range used for the launch of high-altitude balloons and the sounding rockets, which reach heights greater than balloons but lower than satellites. In the 1960s in Kiruna, scientists used to fire these tall, narrow rockets into the aurora and collect particles to see which types of particles, and at which energies, were present during an auroral display. This activity for auroral research continued apace into the 1980s, when it became cheaper to use space-based satellites instead. Rockets still launch, but these days they are more often for experiments on microgravity or atmospheric studies and only occasionally to study the aurora. Now a model of one of the rockets stands in the building of the Swedish Institute of Space Physics. It stretches up within an open foyer from the ground floor to the third floor.

One of the newest and most exciting ventures is Spaceport Sweden, a company pioneering commercial human spaceflight and aiming to establish Kiruna as Europe's gateway to space. Chosen for the combination of high-tech industry and sophisticated research facilities, Kiruna's position in the auroral zone also gives tourists the opportunity to take aeroplane flights to see the northern lights from above the clouds. In the future they hope to offer spaceflights with partners like Virgin Galactic. Future space tourists taking a spaceflight from Spaceport Sweden would ascend into the ionosphere 100 kilometres (60 miles) above sea level. They would experience weightlessness and see the Earth curving away below, the thin blue line of its atmosphere alongside the blackness of the beyond. Perhaps the success of space tourism will be what the new Kiruna needs to launch it into the future.

That night we walked out after 10pm looking for the northern lights, the hair that was escaping my hat beginning to freeze. We didn't see the lights that night. We took photographs in the snow instead, silhouetting the trees against the lights of the city. We found only an orange glow, when we had been hoping for green. In order to see more, we needed to get out of the town.

* * *

WE MET AT the bus stop at 8.30am the next day to travel to Abisko, a scientific research station and prime aurora-viewing destination. Abisko is an hour or so away from Kiruna, set in the mountains towards the Norwegian border.

It seemed colder than the day before as I queued for the bus. As we drove out of town the Sun was rising and the snow around seemed to pinken, highlighting the low mountains in the distance. The trees and bushes were low and spindly, leafless and bent over under the weight of snow, stunted by the lack of light in the Arctic. I watched as the Sun rose higher and the sky became bluer and lighter. Gradually the pink tint on the snow faded and deepened to a purplish grey and the Sun sat some way above the horizon, like a bright bronze coin filling the morning with its orange haze.

The Abisko Scientific Research Station is a cluster of low buildings in an open wilderness area on the edge of the Abisko National Park. It sits by a lake surrounded by low, sub-2000-metre (sub-6560-foot) mountains, the most distinctive feature of which is Lapporten, or the Sami Gate – a smooth, U-shaped valley carved out by ancient ice. The major studies pursued at the station are in meteorology and natural sciences, taking advantage of the abundant and varied flora and fauna in the area due to the ranging topology, geology and climate.

Recent climate change in the region is of particular interest, and during the day we learned about the effects of climatic changes on the Arctic environment. The Research Station was opened in 1912, and so has climate data going back to 1913 in an unbroken record. Every day the scientists collect data from the weather station in exactly the same way that it was collected in 1913. This way, scientists can compare current data with the old data without incorporating any measurement effects. Modern equipment is also used to take new measurements but the scientists continue to collect data the old way too. This provides a unique environmental record.

Their data shows the obvious rise in greenhouse gases and the change in global average surface temperature. But what does this mean for the Arctic? One of the largest effects, as we know, is the melting of sea ice. Huge changes in the ice levels have been seen since 1979 – the extent of the sea ice has almost halved. Naturally, this has implications for rises in the sea level: the ice mass of Greenland alone corresponds to 7 metres (23 feet) of sea-level rise. When the ice melts, the effects will be felt far and wide; low-lying coastal areas all over the world will be swallowed, including a large chunk of Florida, most of the Netherlands and Bangladesh. The water rise will also have an effect on cities such as London, Shanghai, Venice, San Francisco, New Orleans and many more. The ice melting then affects climate change further. White ice and snow has a very high albedo, meaning it reflects a lot of sunlight, between 50 and 80 per cent depending on whether it is melting sea ice or fresh snow. With less ice on the planet to reflect incoming sunlight, more will be absorbed instead. This is just one example of a feedback system that operates to increase the warming further. Tipping points are being approached, and in the Arctic they are already seeing the effects. If the temperatures keep rising as predicted, Abisko will no longer be arctic.

In the early afternoon, as the daylight was beginning to fade, we began our short walk away from the Research Station buildings and towards the nearby mountain. Across an empty road, under a railway line, out we struck into the frozen wilds – over a river buried bank to bank in layers of snow upon snow, rocks visible beneath as pillows of white.

There's a peace about a snowy landscape. Even whilst studying, with full access to the trappings of the modern world, including the Internet, emails and my work, I still felt a peace that can be hard to find at home. Perhaps it is the whiteness, the blank canvas: the purity and the sense of tranquillity that this engenders. Just looking at a tree laden with snow, its branches heavy under the weight, I felt a sense of perfect calm.

On we walked, up a wide, meandering snow path flanked with frozen, filigree trees, until we came upon the old chairlift strung out up the mountain. Two by two, the lift took us up to the Aurora Sky Station. We were given blankets to wrap around ourselves for the journey up, but it was still freezing. Even our eyelashes became white with gentle frost. Before long we rattled into the top level of the chairlift and hopped down onto solid ground again.

The Aurora Sky Station was little more than a large room, though within it there was a warm café and a small aurora exhibition. It was so unremarkable from the outside that as each pair alighted from the chair they headed off further up the hill in search of something else, some other edifice from our imaginations more congruous with its dramatic title. The pointless jaunt up the hill did provide a good viewpoint to watch the sunset, however. The Abisko Scientific Research Station was now in cloud and the mountains enclosed what looked like a large, white lake. The horizon glowed dusky pink and faded through pale orange to blue. The tips of the mountains where the low sun still reached were the same pink hue as the morning.

When we were called back to the Sky Station we found a simple single-storey wooden building perched on the side of the slope and partially buried in snowdrifts. Relieved to get out of the cold, we rushed inside, where we were met with tea and cinnamon buns and a short talk on the northern lights.

It surprised me at first that the explanation of the aurora that we hear most typically is in fact deeply flawed. Often we are taught that the aurora is formed when particles from the Sun are channelled by our magnetic field to the Earth's poles where they interact with our atmosphere. However, this alone cannot explain the bright night displays. Solar particles – predominantly electrons – simply do not have enough energy to cause the vibrant colourful lights, so they must be *accelerated* somehow. Besides, if they are captured by the magnetic field and funnelled directly to Earth they would hit Earth on the side facing the Sun – in other words, during daylight. It is impossible to see the aurora in broad daylight – the auroral light is too weak and diffuse to compete favourably against light pollution from cities, let alone against strong sunlight. Incidentally, that is why we don't see the northern lights in summer. The Arctic is the Land of the Midnight Sun, where the Sun never sets between April and October. The aurora still occurs, but the Sun is so much brighter that it obscures any auroral light. In short, seeing the aurora during daytime is only possible in a few select places during polar night, when it is even dark at midday. These day-side aurorae are very weak, because of the low energy of the incoming solar particles. They are interesting to study, but the direct mechanism of their creation is not the same as that producing the night-side aurora that is most often seen and which we know as 'the aurora'. So then how, you might be wondering, do charged particles get around to the night side of the Earth and create the northern lights there? And how are they accelerated to the higher energies required for those more vivid, colourful displays? That is the magic and mystery of the true aurora, something I will be exploring in this book.

It may seem a minor point, but forgetting the fundamental mechanism of acceleration is not just a simplification of auroral origins, it's a misrepresentation of the facts. It also implies that we have a full understanding of the mechanisms of the aurora, many of which are still under debate. Importantly, it focuses attention on the particles themselves rather than the bigger question of where the energy comes from to generate these awe-inspiring light displays.

It's impossible to say where the electrons causing the aurora come from, apart from that they were bouncing around within the magnetosphere (the Earth's region of influence in space) when something accelerated them to Earth. We can't say that those same electrons we measured coming from the Sun are the ones that have reached Earth and caused the lights. Everything gets mixed up. It is much as Bill Bryson described atoms in *A Short History of Nearly Everything*: 'Every atom you possess has almost certainly passed through several stars and been part of millions of organisms on its way to becoming you. We are each so atomically numerous and so vigorously recycled at death that a significant number of our atoms – up to a billion for each of us, it has been suggested – probably once belonged to Shakespeare.' Could we ever tell which actual atom it was in our body that belonged to Shakespeare? Of course not. Does it matter which? Perhaps not.

Eric Donovan, Professor at the University of Calgary, is astonished that when people talk about the aurora they need to know where the electrons come from. When we flick a light switch, causing a stream of electrons to flow through a circuit and transform their energy into light in a bulb, we don't ask where the electrons came from because it is irrelevant to the mechanism of the transfer of energy. 'The solar wind matters for the aurora because it's where the energy comes from,' says Eric. 'It's how the energy of the solar wind becomes the aurora – that's the interesting bit.'

What we do know is that electrons and protons come from the Sun and electrons are accelerated into the atmosphere where they cause the polar lights. There is still a lot of mystery in the space between these two ends of the process. If we need to generalise, we would be better saying that the *energy* of the solar wind accelerates particles within the Earth's magnetosphere, which then pass into the Earth's atmosphere, not that particles from the Sun are funnelled to Earth.

In reality, scientists at universities and research institutions all over the world are studying the aurora and the connection between the Sun and the Earth in an attempt to understand the complex interactions that lead to this beautiful and fascinating phenomenon. There are indeed still some mysteries hidden behind the writhing, tenuous curtains of the polar lights. This is where my journey will take me.

We didn't stay into the evening at the Auroral Sky Station, though one could. From their vantage point above the clouds the northern lights are seen almost every night. We had to get back to Kiruna, so instead we took the cold journey down in the very last of twilight and walked back to the research station in the dark, guided by small glowing lanterns at the sides of the soft snowy path.

* * *

BACK IN KIRUNA we sought the aurora from ground level. After the first night's light pollution problems we had realised that the best place to view the northern lights was from a ski slope on the edge of town. After nine o'clock, small groups of students could be seen – and heard – marching through the last of the residential streets and out along the dark road towards their best chance of catching a glimpse of the lights.

Up the ski slope the snow was hard underfoot, scraped in pisted lines and frozen in, becoming less firm at the edges where the snow was uncompacted and deeper. No snow had fallen in days and footprints from previous nights were visible, along with scrapes and flattened patches along the centre of the piste where, after the night's viewing, many of us slid down on our bottoms – our sledges large bin liners courtesy of our student accommodation. However, this faster – and more amusing – means of descent, the corduroy ice and the snow shower from heel-braking, made for an uncomfortable ride. As I walked up on the final night I snaked my way up the edge of the piste in a narrow band, not wishing to disturb it any more.

From the top of the slope there were extraordinary views all around. The lights of the town and the mine commanded attention in one direction, while in the other the world was black. Looking north into the darkness, I saw an indistinct, flattened shape –

perhaps a low mountain – but little else. I looked up at the stars: The Pole Star, of course, almost directly overhead; The Plough; Orion; Cassiopeia. The familiar constellations were very visible in the dark sky, and there were many more that I didn't yet recognise. A couple I suspected were planets, perhaps Mars and Venus.

As we watched, I started to put on all the extra layers I carried with me. At -25°C one cools down very quickly. I was wearing thermals and three pairs of trousers whilst on my upper body there were three merino wool thermal layers with a down gilet and a padded ski jacket. I had a Patagonia nano puff too, a synthetic insulation jacket, which I was saving for when I felt even colder later. I was a thick bundle of clothing, with a little face poking out. The atmosphere up there on the hill was fun and convivial, almost like a bonfire party. The students were joking and singing, sometimes jumping around to keep warm.

To the west, the mine looked almost ethereal, dripping white light and with three plumes of smoke billowing into the sky, illuminated from below. I turned my back on the industry and looked north, where a very quiet aurora was beginning to appear as a green, arcing haze. Over time it grew in colour and clarity, the arch becoming more defined, then breaking and twisting into an S-shape. Other parts of the sky were brightening, too, but in different ways. To the right now the light seemed a whiter green, and more diffuse, but had a semblance of linearity, like faint rays. These changes happened so gradually that they were, to begin with, barely perceptible, but if I turned away and looked back I knew a change had taken place.

Over the days when we were in Kiruna the activity level of the aurora was low, and the auroral displays were not classed as extraordinary. We only ever saw green light, for example, no other colours. To achieve a brilliant red or the fabled violets or crimsons, the aurora generally requires more energy. But for me it was an extraordinary experience nonetheless. Despite its quietness, its difference from the photographs that are now taken of the aurora, actually seeing it for myself, in the bitter cold beneath the stars, felt wondrous. Walking beneath the pale green glow and discovering its features, its beautiful twists or drapes, brought not just excitement but gratitude. I said as I flew in that this was only the start of my journey; I knew as I left that the Arctic and the aurora would lure me back for more.

CHAPTER TWO

NORWAY – CULTURE, SPIRITUALITY AND OTHERWORLDLINESS

I BELIEVE YOU CAN get a measure of a place from the feeling it stirs in you in the air. From the air, you already begin to feel the land speak to you.

Imagine being in flight and coming in to land in some particular place. There is a point during the descent where you are low enough, close enough to the ground, to see the colours of its earth, roads, buildings, lakes. You know then whether you feel its character, see its beauty or if it opens its arms to welcome you. Or perhaps, conversely, you feel an aloofness, an appraising appreciation only. As you descend further, skimming the runway, or as you disembark, the horizon shrinks and the viewpoint is narrow. You may enjoy the place, the things you see, the people you meet, but your connection will be made real from the air. It's the ultimate test. It's the way your head becomes light and your breath deep, an involuntary smile creeping to your lips. That is the measure of your love for the land.

It is, of course, a personal and variable measure, often independent of general consensus regarding the 'beauty' of a land. The beauty of landscape is firmly in the eye of the beholder, a beauty wrought of familiarity, time, experience and an essence of one's own character. It is the feeling of home, and it is that tiny element of character that imbues some foreign climes with that very same homeliness. For me, this feeling comes from the higgledy-piggledy fields of Britain from the air; the prominent landmarks of London as the aeroplane cruises west up the Thames towards Heathrow; or from snowy mountains seen from a new level – whole ranges standing unobscured, poised and elegant, their glaciers, ridges and summits open to my scrutiny; and the white, glassy, untouched emptiness of the North.

Having seen the aurora in Sweden, I found myself just wanting to see more; to see it bigger and brighter and more colourful. I realised that seeing the lights *in situ* added something intangible to the experience. I began to think of other, earlier polar explorers and how they might have felt, seeing the aurora for the first time and perhaps having no idea what it was. Often, however many times it is observed, the awe seems not to fade. The Norwegian explorer Fridtjof Nansen seemed to be spellbound and philosophical about the aurora no matter how many times he witnessed it. 'These wonderful night effects are ever new, and never fail to captivate the soul,' he wrote in his diary in March 1894 during the first winter of his expedition on the *Fram*. But these early explorers were not the first people in this hostile, exquisite land deriving comfort from the dancing lights. Human habitation of the North dates back tens of thousands of years, and with Arctic occupation come stories of the aurora. The delicate patterns and striking colours grab people's imagination like no other natural phenomenon. The lights have featured in art, poetry and literature throughout the centuries, and they often have an almost spiritual effect on those they touch.

I decided to go north again, to continue my journey to northern Norway in order to investigate the Sami heritage and learn about the northern lights in folklore. Along the way, I hoped to learn more about polar exploration and the rise of auroral science – the beginnings of our movement from myth to the possibility of knowledge.

Of course, I was looking out the window of the aeroplane as we were nearing Alta – I always want the window seat. The landscape I saw was mostly flat and smooth, the snow as perfect and even as the skin of a china doll. Occasionally, though, blemishes would appear: small creases of winding, frozen rivulets to break the smoothness. In some places the creases and rivulets would get deeper and patches of brown forest shaded the landscape. Creases like wrinkles. Rivulets like scars. Forests like stubble. This landscape stretched on as far as I could see, so intoxicating that it was almost without scale – it could as easily have been a close-up of a patch of windswept snow as a bird's-eye view.

In Alta I met a friend, Cecilie. She lives in Oslo but her family are originally from the far north, from Alta and the fjords around. Cecilie toured me around the very northern county of Norway, Finnmark, so that I might meet her family and the locals and learn what is important to those so far north. The most immediately obvious themes were fishing and dogs.

The first thing Cecilie did was take me to meet dog mushers (drivers) preparing for the Finnmarksløpet, the northernmost sled-dog race in the world. The course crosses Finnmark, stretching for 1000 kilometres (600 miles) up over the northern border of Finland and butting up to the most northwesterly region of Russia. It's a challenging race, where conditions vary from coastal and windy with snowstorms to sheltered and frozen; temperatures can range from just above zero to below -40°C. The organisers describe the terrain as 'mostly barren land', with some birch and pine woodland along the way. Despite the difficulties presented by the landscape, it's a popular race that often attracts over one hundred mushers with over a thousand huskies. For the locals it is a big event. Like the northern lights, the dog race is part of the culture up here, a vestige of the old ways.

Sleds and dogs have long been used by indigenous peoples as a means of transporting food from hunting sites to living sites. Faced with heavy loads to drag, early humans would have looked for a way to reduce the friction between their load and the ground. Reducing the surface area in contact with the ground does just this, so placing the load on two narrow runners aligned with the direction of travel would have significantly eased their burden. The Inuit sled, made entirely of the animals they hunted, had runners made by wrapping fresh fish (arctic char) in wet sealskin and letting it freeze. Later the runners were strengthened with layers of pulverised moss and water, then smoothed with an ice glaze so the friction was minimal.

Indeed, the sled is the earliest form of vehicle and predates the wheel by several thousand years. Moreover, skis would have enabled prehistoric man to hunt more easily in snowy or icy terrain, perhaps even during the last Ice Age twenty-two thousand years ago. The remains of skis and sled runners have been found in peat bogs in Russia dated to almost twelve thousand years ago, and it is likely they were in use before that. Dogs also began to be domesticated around this time, likely deriving from the wolf or the wild dog. This was the beginning of a special relationship between man and dog that endures to this day. In the Arctic, dogs could be used to help with hunting, herding and pulling sleds (and indeed many continue to do so today).

More recently, combining the use of dogs with the use of the ski brought further benefit, and this coupling was employed extensively in polar exploration around the turn of the twentieth century. Dogs could pull greater useful loads without the burden of dragging the men. The interplay of skis and dogs for polar exploration was pioneered by Nansen. Skiing was beginning to gain popularity, and the benefits of this form of travel were clear to Nansen.

Late nineteenth-century Oslo, then Christiania, was the birthplace of modern Nordic skiing. It was then that skiing moved away from being purely for survival and necessity and began to edge into recreation. Early adopters, such as the Norwegian Laurentius Urdahl, ventured into the winter wilderness for fun. Urdahl was a keen sportsman and lover of the great outdoors. He loved exploring the mountains on his skis, carrying everything he needed on his back, rolled up in his reindeer-fur sleeping bag. He and his companion, Bredo Berg, made the first winter ascent of the highest mountain in southern Norway, Gausta, in 1890. At 2000 metres (6500 feet) it is only a modest altitude, but its relatively high latitude of almost 60 degrees north means the conditions and weather are challenging, and indeed they made three attempts to climb the peak. Prior to this, Urdahl had been exploring the Nordmarka – forested terrain near Christiania – and he set up public meetings on ski touring in the city with experts available on the panel for advice and discussion. This new fashion for skiing was buoyed along by 'Nansen fever' in Norway, following Nansen's return from his arduous ski crossing of Greenland in 1889, which was regarded as an heroic achievement, and later his expedition seeking the North Pole on his ship, the *Fram*.

I am fascinated by exploration, particularly polar exploration. I love the stories of people pushing themselves to their limits to increase mankind's knowledge of the world, testing themselves in harsh environments and ultimate survival situations. Within all these severe situations there are moments of beauty – the colours of the ice, the patterns the snow makes in the wind, the call of seabirds or, in winter, the aurora. These delicate lights stirred even those tough men who often wrote rapturous accounts in their diaries. I wanted to learn more about their motivations and ambitions, both personal and scientific to different degrees, and I wanted to gain a better understanding of what they endured in pursuit of those ambitions.

In Oslo I visited the Fram Museum and saw the polar expedition ship that I had read so much about. It rests in a dry dock on the peninsula of Bygdøy, housed in a triangular-tent-shaped building where one can circle the ship whilst reading about polar expeditions past.

The *Fram* was specially designed so that it could rise above the ice and not be crushed as other ships in pursuit of the pole had been. It had a rounded hull so that, as the ice compressed it, the ship would be pushed up rather than crushed, in the way a hazelnut squeezed low between fingers will push up and out of grip. The voyage was unusual in attempting to reach the North Pole by drifting towards it on a ship frozen into the sea ice. Many perceived the idea to be madness. Nansen, however, had studied previous expeditions seeking the pole and had concluded that different routes and methods must be tried. There was a theory of circulating ocean currents in the north polar region, and in 1884 Nansen read that wreckage from an American expedition called the *Jeannette*, which became stuck in the ice north-west of the Bering Straight in 1879, had been found on the south-west coast of Greenland. The vessel had drifted in the ice for two years before it foundered north of Siberia, then continued its progress towards Greenland. Nansen wrote, 'If a floe could drift right across the unknown region, that drift might also be enlisted in the service of exploration – and my plan was laid.'

It was several years before Nansen was able to put this plan into action. In the meantime he completed a doctorate in zoology and undertook the pioneering ski expedition across Greenland. It was 1890 before the expedition of the *Fram* was proposed, funding was secured and preparations began. The *Fram*, christened by Nansen's wife Eva and meaning 'forward', sailed from Oslofjord on Midsummer's Day 1893 and returned in August 1896. It was during the expedition on the *Fram* that skiers and dogs were first used together. Nansen would practise driving the dogs with a sled in the snow-covered ice next to *Fram*, along with companions skiing beside. They saw then that skiers and dogs pulling a load kept up with one another.

In the darkness of winter exercise was difficult and boredom set in, but the men were warm and comfortable. The inside of the *Fram*, though small and low with tight cabins, was attractive in old-style wood and burgundy velveteen. Nansen had ensured the ship was well insulated, installed a windmill to power electric lamps and brought a well-stocked library. There was a semi-automatic organ and other instruments and numerous games. Nonetheless, walking around the *Fram* and considering that this was all that thirteen men had for three years gave a new appreciation of the expedition.

Nansen was a scientist before he was an explorer, and consequently all his expeditions – and others of the *Fram* later – had strong scientific objectives. As well as testing the theory of the east-west current and filling in blank areas on the map, the team made measurements on oceanography, meteorology, marine geology, geomagnetism, flora and fauna, and the aurora, and many of these measurements now constitute important contributions to their fields.

There was much time for reflection. Nansen's diary contains ample philosophising alongside the recording of practical details, as well as descriptions of the northern lights, which they saw whenever the weather was clear: 'In the north are quivering arches of faint aurora, trembling now like awakening longings, but presently, as if at the touch of a magic wand, to storm as streams of light through the dark blue of heaven – never at peace, restless as the very soul of man.'

Nansen and one crew member, Hjalmar Johansen, left the *Fram* in March 1895 in an attempt to reach the North Pole on skis with dog sleds. They didn't make it, but they did get closer than any men previously – 86° 14' N. After a perilous journey back across the ridged ice, kayaking across open water between floes, surviving on little food and fending off a polar bear attack, they reached Franz Josef Land where they over-wintered. In the summer they came upon a British expedition, which they accompanied back to Norway, arriving, coincidentally, five days before the *Fram*, which after almost three years had been released from the grip of the ice.

Nansen recognised the importance of both the skis (which he called snowshoes) and the dogs to polar success. 'Have determined that, beginning from tomorrow, every man is to go out snowshoeing two hours daily,' he wrote in his diary in 1894, the second autumn in the ice. 'If anything obliged us to make our way home over the ice, I'm afraid some of the company would be a terrible hindrance to us, unpractised as they are now . . . If they had to go out on a long course, and without snowshoes, it would all be over with us.' The unpleasant experience of man-hauling on Greenland in 1888 had convinced him to take dogs on his next expedition. He asserts that had he been able to acquire the necessary dogs he would have taken them to Greenland also, believing it, from his study of 'uncivilised polar tribes', to be the safest means of travel over the terrain.

Working together, with men on skis and dogs pulling sleds, the team could move faster. The Norwegian explorers were also immensely pragmatic and ate their dogs as their loads diminished – indeed, they had planned for it. This attitude is perhaps what gave the Norwegians the edge in polar exploration. When Amundsen beat Scott to the South Pole, Amundsen was using skis and dogs.

Most are familiar with the story of the ill-fated English Captain Robert Falcon Scott. His team had immature motor sleds, which broke down almost immediately, ten ponies, which were not suited to the terrain or conditions, and half as many dogs as Amundsen, which were sent back to base after a month due to lack of food for them. Scott arrived at the South Pole in January 1912, after seventy-nine days of gruelling trudge, to find a small tent flying the Norwegian flag. Another team had beaten them to it. They never made it home.

The team who made it to the pole first was headed by Roald Amundsen (erstwhile protégé of Urdahl) who sailed down to Antarctica in the *Fram*. Amundsen himself put the team's success down to the use of skis. They were able to move quickly and efficiently over the snow, conserving strength and keeping morale high. They didn't have to suffer. They were making an average speed of 24 kilometres (15 miles) per day to Scott's 17 kilometres (10.5 miles). Scott's party also took skis for this, his second expedition to reach the South Pole. However, as recent converts to the merits of the ski, the British were rather incompetent in their use, so skiing was often more tiring than walking. They struggled on exhausted. The Norwegians, by contrast, had ample skiing experience. Since the first ski tour that Amundsen had done with Urdahl – an unsuccessful crossing of the Hardangervidda at Christmas 1893, aged just 21 – Amundsen had continued to ski in Norway and further afield in the Canadian Arctic and Alaska, skiing in the North American forest as well as on the tundra. He was in fact the second man in history to reach the North Magnetic Pole and he did so on skis. The different conditions he encountered, like the loose forest snow rather than the hardened, frozen-over snow of the treeless tundra, improved not just his skiing but his mastery of the environment. He learned about Inuit fur clothing that he could wear whilst skiing to keep warm without sweating. He understood the advantages and limitations of his skis and sledges and was able to make adaptations and new developments to skis and boots to improve their performance. Importantly, by the time Amundsen was skiing in polar regions he had already learnt some difficult lessons at home in Norway. As Roland Huntford, author of *Two Planks and a Passion (The dramatic history of skiing)*, puts it, paraphrasing Urdahl, 'Amundsen had committed his beginner's follies where they would not be fatal.' Scott was perhaps not so fortunate in his heritage.

* * *

THE MORNING AFTER my arrival in Alta, Cecilie and I walked into the town centre to see the start of the Finnmarksløpet dog race. It was sunny and the centre of town had rather a festive feel. There were flags in red, white and blue lining the route (and blocking the view of many of the spectators who were stood back from the barrier), and music blared as a man on a loudspeaker entertained the crowd. We found ourselves a cramped space by the barrier and waited for the race to start. It wasn't long before the first musher took his place at the start line.

The dog teams were set off one by one, every minute or two.

Each musher received a short introduction from the man on the loudspeaker as the dogs were readied. There was much barking and howling, the dogs jumping around excitedly, seemingly eager to get going. The commentator gave a 5-point countdown, somewhat arbitrarily in either Norwegian or English, to send them off. After all the jumping and fidgeting, once the dogs were set going they all settled into line and rhythm, their little sock-covered feet padding the snow, the musher hanging on the back, standing on the narrow runners and waving to the crowd. We watched almost seventy out of about 140 teams go by to start. Then we decided that we had seen enough and went for lunch.

That afternoon we went on snow scooters to one of the dog-race checkpoints, many of which are inaccessible from the road. We drove about half an hour out of Alta and met two of Cecilie's friends who each had scooters. It was my first time on a snowmobile and Cecilie told me to ride with her friend Kenneth. I put on my balaclava and goggles to protect me from the wind and gingerly climbed on behind him, feeling a little apprehensive. Initially I held lightly to his waist, gripping the snow scooter tightly with my legs. It was scary at first; I have to admit I squealed a couple of times when Kenneth accelerated and the front of the scooter began to lift off the ground. We were skimming across a wide, frozen lake, but despite the apparent flatness I felt every bump. We were going fast. Kenneth turned and shouted to me to 'stick to my back.' I held onto him tighter. Each time we went over a bump and lifted off the seat together we got closer, so gradually my legs were even a little under his. It felt more stable like that.

We reached the checkpoint and stopped for a while. As I got off the scooter my legs felt weak from gripping so hard. I removed my goggles and looked around. We were by a small cluster of houses in a clearing beside the lake, surrounded by low, scrubby brush. Here officials stood around in fluorescent green bibs and the track was marked out with wooden stakes, but people could cross it as they pleased. There were barely any people there to watch, unlike the crowds in central Alta that morning. Whenever a team came through, the musher would stop the dogs and the green bibs would crowd in to talk while the dogs shook and nuzzled.

As the afternoon went on, the clouds were coming in and it was getting windy. The weather could change quickly; inland the climate is more Arctic than at the protected fjords. At least on the return ride I felt much more relaxed. I tucked in closer to Kenneth this time and linked my fingers – in my huge, red down mittens – in front of his body, so I knew I wouldn't fall off. This time he went even faster. Sometimes he went so fast that the front of the snow scooter raised up high in the air and we skimmed along for a few seconds on the back of it, like a giant bike doing a wheelie. By this time it was snowing, the light was fading and the cloud was low, so visibility was worsening. Bright snowflakes glittered in the headlights. We travelled back across the frozen lake, wide and barren. I could just see the hills rising up at the edge of the lake, and the small bushes and branches pushing out through the snow like the stubble I had seen from the air. We were travelling alongside the dog tracks and each time we passed a dog team Kenneth would wave. I would not let go of him, so for me waving was definitely out. As we passed through the fog patches visibility got increasingly bad and all I could see were the looming headlights of other approaching scooters. The noise of the huge engine was in my ears, the force of the wind in my face.

So we skimmed across the lake, two tight figures on a speeding hulk of plastic, metal and petrol, so close that he was almost sitting on my upper thighs, my face buried into his neck. This could be quite sexy with a boyfriend or husband, I thought to myself. Maybe that's why the Norwegians like it so much. Snowmobiling is very popular; Kenneth told me that most Norwegians have a cabin somewhere in the mountains, and there they will have a snowmobile, probably one each. For days out they will wrap the kids up in a skidoo trailer, pack a picnic and head out into the wilderness. For fun or practice children can be towed along on mini scooters behind a parent on the skidoo, rather like water-skiing. I saw a father with two children being towed next to each other on separate mini scooters. So they start young. Kenneth had been on scooters all his life. He began driving at 11, though now the legal age is 16. 'It wasn't so important back then,' he said.

* * *

'You should have been here earlier in the winter. When it was colder,' said Knut, the Sami reindeer herder whom Cecilie and I had joined for the day.

'We haven't seen them at all since I've been here,' I replied. 'That's nearly a week. It's been cloudy every day.'

He nodded, explaining that the coldest months, January and February, were the best to see the lights. I was starting to feel that it was my fate never to see the aurora in full flourish.

We were out in the hills somewhere around Karasjok, a few hours' drive inland from Alta near the Norwegian-Finnish border. Karasjok is the Sami capital in Finnmark, home of the Sami parliament and cultural collections, which we had toured the previous day. We were staying in a beautiful husky lodge where our personal hut and almost everything inside was made from slate, wood and reindeer hides. Old, stripped branches formed coat or towel hooks; larger logs and ropes came together to form inventively designed shelves in the kitchen; the kitchen worktop was a huge piece of thick grey slate. A vast ironwork candelabra hung above the large wooden table, flanked on one side by a sofa strewn with reindeer hides and on the other by some small stools. One wall was entirely made of glass and looked out onto the snowy pine woodland of the husky lodge site. Other small wooden huts were just visible through the trees. A short tramp through this woodland had brought us to the wooden cabin of the lodge owner, Sven Engholm, and beyond that the large pen of the numerous husky dogs chained next to their kennels. In the dim, wooden kitchen of the cabin,

Christel, tall and thin with long dusky blonde hair plaited down to her waist, told us of the Sami man who could take us out to see the reindeer herd.

So there we were, out in the snow and sunshine on the Finnmark plateau, Finnmarksvidda, wrapped up to our ears in big down jackets with hats and mitts. Knut seemed to be quite warm enough in what looked like a biker jacket. We had driven out for almost an hour from the lodge to find the herd. Cecilie and I were tucked up in a rudimentary wooden sled which was being towed behind the snowmobile. It looked homemade and was painted a dark brownish-orange. It was a pleasant ride, and we passed first through the pine forest near the lodge, travelling on narrow paths only big enough for the snowmobile and sled. We travelled quickly, the trees and naked twiggy bushes rushing past beside us. The rush of the wind was in our ears, the overtone against a background of engine whirring and the creaking and clattering of the tow coupling. Soon we were out of the forest and skimming quickly along the Karasjoka River, wide and flat like an empty, snow-covered motorway. Later we were back onto the gentle winding tracks in a woodland of slender silver birch. These tracks are maintained, marked and – as I found out – policed. Out there in the wilderness, having passed not a single other person on our journey thus far, we met two policemen in high-visibility, fluorescent-green, banded vests. They drove white skidoos on which 'POLITI' stood out in shining blue above a red stripe. We stopped and there ensued some serious conversation in brusque Norwegian until all was resolved. Then the tone became jovial and the police wished us well, posing for a photo with Cecilie before moving on. It turned out that our driver had received a warning for not carrying his snowmobile licence with him – these are serious roads after all!

Eventually the wood thinned out to a relative clearing scattered with spindly birch and crisscrossed with deep footprints. We had found the herd.

Knut stopped the snowmobile and got off. From the sled behind us he took out some straw which he threw around liberally a fairly short distance away – maybe as little as 20 metres (65 feet) – to tempt the reindeer nearer. We could hear the tinkling of bells as they approached. Not all the animals have bells, just the older, more experienced reindeer which know the area. Younger reindeer will follow the sound. 'They are my little helpers,' said Knut.

As Cecilie and I quietly watched and photographed the reindeer, Knut made a large fire in a pit of squashed snow and laid out reindeer hides so we could sit. He built the fire directly onto the snow, piling the wood high. Without petrol and without even any obvious kindling, the fire blazed happily for the hour and a half that we sat and talked. Knut had a black metal kettle that he set on top of the fire, hanging it from a birch stick that he stuck into the snow bank behind. He made coffee, pouring the instant granules into the pot and letting it heat. That was when he told me it is better to be there when it is colder, in order to have more chance of seeing the aurora. When the coffee was hot he poured it into small mugs carved from the ubiquitous birch and passed them over to us. As we settled down, I asked him about his experiences of the northern lights.

'They used to scare me, the northern lights,' he began in his gruff, accented English. 'My parents told me if I made the lights angry they would swoop down and take me. Yes,' he mused as he rolled himself a cigarette, 'the northern lights were like a natural babysitter for us. We were always running home before it got dark.'

There is a famous Sami story about making fun of the northern lights. Two brothers were travelling up in the mountains to collect their reindeer herds. One of the brothers, Biete, began singing made-up songs disparaging the Sun, the Moon and then the stars. Each time, his brother Garrel tried to make him stop, telling him evil would befall him if he made fun of the heavenly creations. But Biete would not listen. Then, when he made fun of the stars a fork of lightning flashed down and killed the reindeer pulling Biete's sled. His brother again urged him to stop, pointing out the lightning flash as evidence of the heavens' anger. Still Biete continued. That night, when he sang his teasing song to the northern lights, they began to streak wildly across the sky. The northern lights struck Biete and killed him. Garrel, unsurprised, simply drove away sadly.

We southern travellers enjoy the beauty without the fear, unencumbered by the legends, but their myth has a certain menace.

'So you didn't go outside to watch?' I asked, surprised.

'No, we were hiding behind the curtains!' Knut laughed. Then he added, 'If we were many together it wasn't so dangerous, wasn't so frightening. Then we would watch.'

'What did your parents tell you that it was?'

'It was the spirits of the dead people, our relatives. They were watching us.'

I asked him if he remembered his best view of the lights, but he could not: he had seen them so often he couldn't separate them in his mind. 'The northern lights is so common to us,' he told me, 'it is there like the clouds and everything else up in the sky.' But things were gradually changing. Global warming and the rising temperatures meant that the lights were appearing less frequently, he said. Could there really be a connection? Perhaps without the cold weather they don't get so many clear nights, I mused.

The fire crackled and we sipped our coffee. The reindeer bells tinkled in the background.

We went on to talk about the colours of the aurora and Knut said that they see more colours when the weather is changing. They don't use it as a marker of weather – Sami in the past used the behaviour of animals and the colours of the sunset for that, now they can use the weather forecast – but he says that if he knows the weather is changing he sees different colours, maybe more red than blue. I wonder about this, since I know the aurora occurs much higher up in the atmosphere than the weather we normally experience, but if Knut has seen them a lot he has had time to notice patterns.

Suddenly the wind changed and the smoke blew steadily right into my face. Cecilie jumped up spluttering and I shifted position, trying to get out of the direct stream. Knut laughed and said, 'In Sami, if the smoke goes towards you then you are in love!' He chuckled again.

'Oh really?' I exclaimed. 'I'm in love! But who with?'

'I don't know,' he said, '*you* have to know.'

Neither he nor Cecilie were to be my perfect match, though I took it as a hopeful sign.

By this time the fire had sunk into the snow, but the remnants still burned, splattering the surrounding white snow with black flecks.

Knut is one of several generations of reindeer herders. His parents passed down the knowledge to him, as did his grandparents before them. But it is not the same as it once was – technology has changed things. The Sami now live in towns, the children go to school and the women work. His parents were the same. It has been over the last two generations that the Sami have settled. Knut grew up in Karasjok with settled parents, but in his grandparents' day the Sami were on the move.

'When my mum was a little girl she had to live in the school in winter,' Knut told us, 'because my grandparents were away with the reindeer herd.'

Now, with snowmobiles, even the herders can live in town. They drive out every morning to gather the reindeer, make sure they are keeping away from other herds, and to move them if the snow becomes too deep. By snowmobile it usually only takes around thirty to forty minutes to reach the herd; in the old days they had to go on skis or with a reindeer to pull a sled. So the nomadic people settle. Knut says they can no longer live by herding alone so the other family members must also get a job. Herding has become 'an expensive hobby'.

In the summer the herds move north and the Sami follow. They move to their summer area with some other families, staying in small wooden houses there that are left shut up in the winter. Cecilie and I had passed some earlier in the week as we drove north from Alta towards Hammerfest. The low buildings stood in small, desolate clusters in the wide white expanse, looking wind-battered with their roofs almost swept of snow and high drifts piled up against the doors.

Knut explained that, because of the mosquitoes, he needed to move his reindeer in the summer. 'We have to move to the coast or up into the hills where it is more windy and cold. The reindeer, they will start walking north by themselves in April to give birth to the calves. Then I have to follow them. My family will come too. It's a tradition that in springtime we all go together.'

The calves are born in May or June, and in September Knut marks them by cutting their ears. The mothers only have one calf each – more than that and the calves wouldn't survive. 'Every year in May there comes a blizzard, just to kill the ones that are weak. It's natural.' They don't try to save the weak ones.

As for other technologies, these tend to be adopted with varied zeal depending on the generation, or perhaps the temperament of the individual. 'The snow scooter is enough technology for me,' says Knut. 'I don't like the electronics. I can't use a smartphone, it's too difficult. I did buy one smartphone. I went up into the mountains when it was 40 below and it didn't work. So I gave it away.'

I suggested that it might be hard to reconcile the old and new ways of life, and Knut agreed. 'Not for me, but for when I'm trying to teach my children. They are fifteen and twelve. They sit on the computer 24/7. They have no interest in these things, in the reindeer.' He paused and emptied the dregs of his second cup of coffee. 'When they were small then they would be with me every day, when it was such a job to have them with me. But now they won't come.'

'Teenagers,' nodded Cecilie knowingly.

'I try to teach them everything,' Knut continued, 'but I have two girls so they won't be reindeer herding.' One day Knut will have to sell the herd, unless he can pass it on to his nephew or if one of his girls finds 'a good boy', but he acknowledges that it's much easier to learn the practice from childhood. He doesn't know what will happen to his herd in the future.

As the fire burned low and we finished the biscuits, Knut told us how he believes the modern world is killing the old stories too, like the tales of the northern lights. 'These days children don't believe the stories. They see much worse things on telly or on computers. So the natural babysitters are gone. No more sea monsters, no more spirits of the northern lights.'

It is a similar tale with other indigenous cultures of the north. They absorb new useful things into their lives whilst trying to retain their culture, their language, their identity. The Dene* people, from the Northwest Territories in Canada, whom I would visit later in the year, have also embraced the skidoo to enable them to trap from home rather than travelling the trap circuit, rather like the Sami use the skidoo to herd from home. They also use communications like VHF radio, mobile phones and satellite phones. Mike Mitchell, whilst giving me a tour of the Prince of Wales Northern Heritage Centre in Yellowknife, Canada, told me a story that demonstrated what he called the 'cross-pollination of technologies'.

The story was about the Mountain Dene and their traditional moose-skin boat. During the heyday of the fur trade in the eighteenth and nineteenth centuries, the Dene would build these large boats to transport furs (and meat), procured during winter trapping in the mountains, back downriver to the fur trading post. Now this traditional technology is in danger of dying out. The issue was documented in a film called *The Last Mooseskin Boat*, which was made in 1982, in which an elder who remembered how to make the boat from his youth taught the younger ones. Now, thirty years on, the young ones are now elders, and they recently taught a whole

---

* Pronounced 'den-nay'

new generation to make another moose-skin boat. Together they went up to the mountains, built the boat and then sailed it back down the river to their community. What was interesting was that by this time several of the communities had cellular service and many of the elders had mobile phones. So there they were, coming down the mountain river, and from the boat the elders were filming and talking on their phones with their friends in the community. It seems they absorb some of the helpful, modern technologies but try to keep the old ideas and practices alive. It can be a tough battle sometimes.

But what of their folklore? Could it be that they are losing the stories, too? John MacDonald, author of *The Arctic Sky* and someone who lived among the Inuit in the Northwest Territories for more than a decade, thinks they are. He argues that, as cultures modernise, Inuit star lore 'has suffered a rapid dilution', even facing imminent loss. People are losing time for stargazing and stories as lives change and move faster, with fixed housing and snowmobiles. The Inuit and Sami are no longer travelling together as families, sharing their folklore on dark winter nights or long dog-sled journeys. Light pollution in communities can make it harder to see the night sky. Of the elders MacDonald interviewed, many insisted that their knowledge of astronomy was greatly lacking compared with that of their parents or grandparents. 'Indeed,' he says, 'to many, the names of particular stars had been forgotten, their exact location in the sky uncertain, and the legends and narratives associated with them often remembered as fragments of a more complete story heard long ago.'

* * *

SOMETHING THAT FASCINATES me about the aurora is the feeling of spirituality it engenders, regardless, it seems, of age, creed, belief or indeed the epoch in which one lives. Even now that we understand more about the origins of the aurora, that does not dispel the magic of the display, and people still flock to see it. Indeed, many also experience that feeling of spirituality that is the root of the folkloric stories of indigenous peoples, perhaps even in the roots of religion.

Aurorae have been occurring on Earth since there was first a magnetic field and an atmosphere, this combination having endured for at least 80 per cent of the Earth's 4.6-billion-year history. The Earth has had a magnetic field for most, if not all, of its life. The atmosphere was created at least 3.8 billion years ago by volcanoes releasing gas as the young Earth cooled. This was not the atmosphere that we now know; the early atmosphere was devoid of oxygen, which was created by early anaerobic life forms only about 1 billion years ago.

Our ancestors, *Homo erectus*, first appeared in Africa about 1.8 million years ago, and the earliest form of anatomically modern human (*Homo sapiens*) only appeared around 100,000 years ago. They moved into Europe about 40,000 years ago and over the land bridge to America, it is thought, at the end of the Ice Age around 15,000 years ago. By this time, modern humans had occupied regions from the tropics to the Arctic, adapting to harsh deserts and frozen tundra, using their intelligence to develop technologies and foraging strategies to cope with extreme conditions. Besides that, they had begun to develop religious and spiritual beliefs.

Unlike other natural phenomena, the appearance of the aurora predominantly in northern, hostile locations meant that relatively few humans would ever have had the chance to gaze on the spectacle, so early writings on the aurora can be hard to come by. Occasionally, though, large auroral displays push south and are seen at much lower latitudes than normal. Perhaps the earliest indication of the human appreciation of the aurora comes from cave paintings in Rouffignac, south-west France, dating back almost ten thousand years. It is said that on the red clay ceiling of the cave there are lines depicting the curtain-like shapes of the aurora.

The earliest written account of an aurora, if it may be taken as such, could be found in the Book of Genesis in the Old Testament of the Bible, which goes back to the second millennium BCE. Genesis 15:17 reads: 'And it came to pass, that, when the sun went down, and it was dark, behold a smoking furnace, and a burning lamp that passed between those pieces.' The furnace and the lamp (later translations speak of a 'blazing torch') represent God, who is portrayed as heavenly light, but the imagery is exactly that of later descriptions of the northern lights. Further possible auroral references are found in Jeremiah and Ezekiel. Jeremiah tells of 'a boiling pot, tilting away from the north,' which has some parallels with indigenous Arctic folklore. Ezekiel describes 'visions of God', going on to write that 'a whirlwind came out of the north, a great cloud, and a fire unfolding itself, and a brightness was about it.' This seems a fairly clear description of an auroral display. It is understandable that moving, changing forms, seen on the rarest of occasions, could be attributed to visions of a heavenly nature.

Ancient China also offers us some early references. There is a story from before 2000 BCE of the mother of the Yellow Emperor Xuan-Yuan who saw 'a big lightning around the Su star of Bei-Dou (Ursa Major α) with the light shining all over the field.' The story goes that she then became pregnant. Robert. H. Eather, in his book *Majestic Lights*, attributes the auroral interpretation of this story to three main facts: the star described was in the north; there was a lot of light; if the phenomenon was real lightning during a thunderstorm then the clouds would have obscured the stars. He adds that, in those times, rare sky happenings were often linked with the births of important people.

Moving on in time to the fourth and fifth centuries BCE, we find the northern lights in Greek and Roman literature. Indeed, Hippocrates and his student even proposed a theory that the aurora was caused by reflected sunlight. Aristotle also wrote about the aurora and put forward his own ideas for its creation based on his elemental science – fire, air, water and earth. He theorised that water vapour that evaporated from the ground by the heat of the sun collided with the fire element and burst into flames, the light scattering through the air below and producing the display. After this period there was very little thought put into the origins of the aurora until the Renaissance.

Many writings on the northern lights seen at temperate latitudes describe some sense of foreboding or see the display as a portent or warning, usually with a military significance. This may be due to their colour as well as the rarity of the sight. Lower-latitude aurorae are redder in colour, rather than the typically green Arctic lights, and so they were sometimes mistaken for fire. The red colour was also seen as representing blood in the sky, therefore they were interpreted as a bad omen. Rayed structure was often reminiscent of military spears; arcs and curves as fiery dragons. Aurorae are said to have foretold the death of Julius Caesar (44 BCE) and presaged the American Civil War (1860).

Whilst not necessarily seen as portents by more northern inhabitants, except perhaps the very great displays, many indigenous people were respectful if not fearful of the lights. However, here the northern lights were a more familiar occurrence so they were naturally woven more tightly into their culture. Broadly, the stories we hear can be grouped into Reflections or Spirits, except for a Finnish legend that sits apart. This tells how the aurora is caused by an arctic fox running and striking fire in the snow with his tail. The Finnish word for aurora even comes from this story: '*revontulet*', meaning fox-fire.

In some cultures aurorae are interpreted as reflections. Eather describes an interpretation from Norway as 'reflections of schools of silver herring in distant oceans.' In Danish folklore, the northern lights are reflections of racing swans trapped in the ice, their flapping wings creating the flickering lights of the aurora. Old Norse mythology speaks of the aurora as reflections of light from the shields of the Valkyries, the beautiful virgin warriors who rode over the battlefields choosing the slain. This last story crosses over into the domain of the dead and so shares some common ground with the folklore of spirits.

Knut, the Sami reindeer herder, said that his parents told him that the northern lights were the spirits of their relatives. This is a common description between various, well-spread groups – the spirits are the common thread, but their activities change. The Inuit of the Hudson Bay area of Canada tell of spirits feasting and playing football with a walrus skull. Inuit in Greenland talk of 'the ball game of the departed souls'. Brilliantly, the Inuit word for aurora, '*aksarnirq*' means 'ball player'. The Inuit also described the 'whistling, rustling, crackling' sound of the aurora, which is only sometimes reported and is still something of a mystery scientifically, but is probably attributable to small electrical discharges.

In East Greenland the aurorae are the spirits of stillborn children dancing with their afterbirth. The Chukchee of Siberia and the Russian Sami tell of souls who died violent deaths. The Mandan Native Americans of North Dakota saw the lights as fires where medicine men stirred great simmering pots (often said to contain their enemies!). This description of the northern lights as a steaming pot resonates with the biblical story of the boiling pot in Jeremiah, giving more weight, one might say, to its interpretation as a description of the aurora. The myths are rich and diverse.

As part of the fabric of northern lives, the aurora borealis features in folk tales and legends, sometimes in a starring role, sometimes as a by-the-by. The lights are part of the landscape. One of my own favourite stories is a Siberian folk tale about the daughter of the Moon and the son of the Sun. The Sun wishes his son, Peivalke, to marry the daughter of the Moon, Niekia. The Moon refuses, wishing Niekia instead to marry Nainas, the Northern Lights, when she is older. To protect her from the Sun, the Moon places Niekia with an old couple living alone on a frozen island. Later, Niekia and Nainas meet, fall in love and marry, but Nainas must leave every day to do battle in the skies across the ocean. Should he stay, he would be pierced by shafts of fire from the Sun. One day, lonely Niekia dupes Nainas into staying late into the morning. She had secretly embroidered a reindeer hide quilt with stars and the Milky Way in beads and silver thread, and she had hung it above him on the ceiling of the hut. Nainas did not realise it was morning until too late. When he did, he flew out of the hut into the day, but the Sun pierced him with fire. Niekia ran to him and shielded Nainas from the Sun with her own body. He escaped to safety, but the Sun, furious that she would never marry Peivalke, flung Niekia back up into the arms of her mother the Moon, where her face can be seen to this day, looking down over the sky battles of the Northern Lights.

Many stories, from different groups of people and spread all around the Arctic Circle, have that same root in spirits. The Norwegian ethnologist Odd Nordland proposed this spiritual connection as evidence of migration paths of the past, the people spreading their stories with them. But could it be that all people feel the spirituality? Certainly polar explorers have done so. Nansen writes that it is 'as though one heard the sigh of a departing spirit.' Scott, on his fateful expedition to the South Pole, wrote of the southern lights, 'it is the language of mystic signs and portents – the inspiration of the gods – wholly spiritual – divine signalling.' Even modern-day explorers like Felicity Aston, whom I met later in Iceland, say that being aware of the origins of the aurora cannot diminish their effect: 'It feels so otherworldly, even when you know the science. That's why people still love them.'

With the north now open to tourism much more than it was even a hundred years ago, we are all able to share in the wonder. The impression is still the same. Kerensa Jennings, a colleague who saw fantastic northern lights displays in Iceland, talks of 'witnessing something both mystical and magical.' She said of the moment, 'it's the closest thing I have ever experienced to something truly spiritual. You got the sense that a higher being, a higher order of some sort, must be present because there was too much activity, too much energy, for it to be spontaneous.'

Witnessing the aurora for myself, it's easy to see the provenance of the spirituality in the old beliefs. Despite knowing more about the science, we feel it even now. A hundred years ago, the aurora was still a mystery to science. Even Scott asked, 'Might not the inhabitants of some other world (Mars) controlling mighty forces thus surround our globe with fiery symbols, a golden writing which we have not the key to decipher?' Scott was partly right about the ethereal qualities; however, maybe now we do have the key to decipher the rest.

* * *

FEAR AND SUPERSTITION surrounding the lights continued into the early seventeenth century, so despite the advances in science and art during the Renaissance very few people made any attempt to describe the nature of the aurora. Those who did were inspired by several strong northern lights displays seen in Europe during the first quarter of the century. The eminent astronomers Kepler and Galileo, and the French mathematician Gassendi, wrote accounts of aurorae in Europe in 1607 and 1619, contemplating their character or origin. Galileo, credited with coining the name aurora borealis, had a theory that there is very high and unusually rarefied 'vapour-laden air' surrounding the Earth. He surmised that 'its upper parts are struck by the sun and made able to reflect its splendour to us, thus forming for us this northern dawn.'

This early interest faltered from around 1620, when solar activity dropped to an uncommonly low level and auroral displays in Europe became a rarity. This persisted for the next hundred years, with only a short interlude in around 1661 which provided some auroral displays. Even in Scandinavia displays were infrequent. Physicists also noted that very few sunspots were seen on the surface of the Sun at this time either. This low-activity period became known as the Maunder Minimum, named after the superintendent of the Solar Department of Greenwich Observatory.

Activity restarted in the early eighteenth century with a spectacular auroral display observed over Europe in 1716. This was witnessed by Edmond Halley (he of comet fame) who had been reading avidly about the aurora and theorising awhile despite, until then, never having seen it himself. His theory entailed 'magnetised effluvia' . . . a substance that could pass through solid bodies. This, he said, could pass through the Earth and out again, carrying with it something that would produce the light. Although his theories were somewhat far from reality, he did envisage this magnetic substance circulating along magnetic field lines, which does contain an element of truth.

From then on, aurorae were again seen fairly regularly and now people began searching for a physical explanation. Scientists and thinkers offered various theories. There was discussion of 'vapours'; of an association with earthquakes; of the aurora being the reflection of sunlight from polar ice; or that large air flows to the polar regions in autumn and winter could catch fire. All these ideas were offered as explanations for the northern lights. At this time strong similarities could be seen with northern descriptions and stories; reflections and fire were very much part of indigenous folklore, and scientific knowledge was not yet developed enough for new ideas to break away.

The first suggestion of a connection between sunspots – activity occurring on the Sun – and the aurora was made in 1733 by Jean Jacques d'Ortous de Mairan, a French geophysicist, astronomer and chronobiologist (whose work would later inspire the field of circadian rhythms). He had his own theories on the aurora and even published a textbook on the subject, possibly the first ever dedicated to the phenomenon. However, his work was not widely accepted, and as interest in the aurora intensified, various arguments about its nature began between scientists.

The mid-eighteenth century saw the discovery of electricity and the invention of the discharge tube. A glass tube with an electric connection at each end was filled with low-pressure gas, then the ends of the tube were connected to a high-voltage electrical supply. The gas inside would glow with a diffuse light, the colour varying depending on the gas inside. Scientists began to agree that electricity was behind the aurora, then in 1741 the Swedish scientist Anders Celsius and his student noticed that during auroral displays a compass needle was affected. (This may have already been noticed by Russian sailors to Svalbard, but the experiences of sailors and fishermen, who often ventured out ahead of the now-celebrated 'explorers', were not usually considered of value to the scientific establishment.) It was clear from the wavering of the compass that there was a connection between the aurora and magnetic disturbances, as well as something electrical. These would prove to be breakthroughs in auroral study.

By the end of the century many scientists were making observations of aurorae and in particular attempting to determine their height. Halley had suggested using the old method of triangulation years earlier, and the French scientist Jean-Jacques d'Ortous de Mairan had also tried the same approach. This involved making observations of the object to be measured from two places of known separation tens of kilometres apart using a theodolite. This is an instrument that measures angles, and it is mounted such that it can move in both the vertical and the horizontal planes to measure both vertical and horizontal angles.

Imagine holding your arms out in front of you so that one is parallel to the ground and the other points to something on the wall. The angle between your arms is the vertical angle. Point one arm in the direction of the object and the other towards a friend – point B to your point A – and the angle between them is the horizontal angle. Knowing the angles from two different observation points and the separation between them, high-school trigonometry can be used to calculate the height of the distant object. This method was used to determine the elevation of the highest mountains in the world. The Great Trigonometric Survey was started in India in 1802, and over the following decades the skyline of the Himalaya was mapped. The altitude of Mount Everest (Peak

XV in the survey) was declared at 8840 metres (29,002 feet) in 1856 after several years of calculations. The most recent measurements suggest its actual height is 8850 metres (29,035 feet).

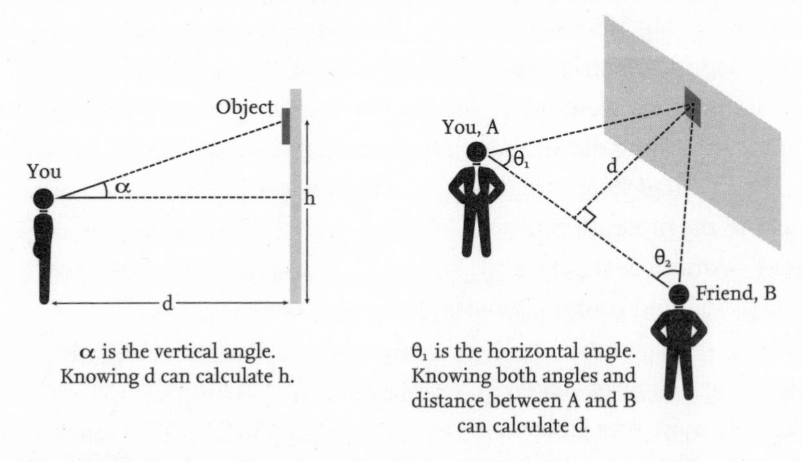

The height of a distant object can be found using multiple angles, distances and trigonometry.

However, determining the height of a static mountain was quite straightforward in comparison to measuring the height of the aurora, the shape of which was always changing. In a time before telecommunications, observers couldn't be sure that they were measuring the same thing – that they were measuring to exactly the same point at exactly the same time. Consequently, estimates of auroral height ranged from within a few kilometres of the ground to a few thousand kilometres into the air. It would have to wait until the early twentieth century – over 150 years – before, aided by telephones and photography, the height could be correctly estimated to between 100 and 300 kilometres (62 and 186 miles). Green light is produced lower in the atmosphere, at around 100 kilometres (62 miles), while red is seen higher, at 200–300 kilometres (124–186 miles). This determination of height put paid to the old folklore tales of aurora that could come down and 'get you' or burn your hair. The northern lights were far too high for that – even the clouds didn't get close.

By the early nineteenth century the state of auroral science was thus: electricity was the preferred mechanism but its role was still uncertain because nobody had been able to observe anything specifically electrical. The height was still in question. However, the other, old, ideas still abounded – the vapours, reflection, combustion, electric discharge and cosmic dust. At this time, auroral studies were advanced by exploratory polar expeditions that were sailing out in search of a Northwest Passage.

People had been dreaming of a Northwest Passage since the close of the fifteenth century when Christopher Columbus "discovered" America (1492) and John Cabot sailed from Bristol to Newfoundland (1497). They sought an alternative trade route from Europe to the riches of the Orient, without having to navigate around the Cape of Good Hope. There was speculation that around the pole there was an open ocean and plain sailing if one could get through the ice at the Arctic Circle. Expeditions found funding either from investors hoping to get rich from mineral exploitation, furs or spices, or by national scientific bodies hoping to extend geographical and scientific knowledge. Exploration flourished, particularly in England under Queen Elizabeth I, and ships pushed further north over the following four hundred years, investigating, surveying and naming islands and waterways after the explorers, the investors or the monarchy. The Arctic was a geographical puzzle that was pieced together over centuries.

Come the early nineteenth century, both the north coast of Russia and the north coast of America had been mapped and it was clear that there was no Northwest Passage south of 64°N*. Explorers would have to venture further north to find a route

---

* Incidentally, it was Roald Amundsen who was the first successfully to traverse the entire Northwest Passage in the ship *Gjøa*, during the same expedition in which he reached the North Magnetic Pole and spent two years on King William Island learning about the Inuit, 1903–6. The passage is entirely within the Arctic Circle, above 66.5°N, and is frozen over for most of the year.

through, where the changing ice blocked their passage, the winters were long and dark and the aurora played in the skies.

Auroral observations made during these polar expeditions led to the realisation that auroral frequency increases as one travels north, but does not keep on increasing all the way to the poles. There is a region – an auroral oval – where the northern lights are seen most often. Scientists also became aware that auroral displays happen in the north and south simultaneously, at opposite ends of the same magnetic field line. Highly descriptive accounts of the aurora written by explorers increased the public interest and scientific theories continued to be debated. Maps were made of the geomagnetic field using magnetic data supplied by seafarers. By the mid-nineteenth century, people began looking for an explanation outside the sphere of the Earth. In 1856 a professor at Yale University, in America, Denison Olmsted, proposed that the aurora was 'cosmical in origin'. After a large – and now famous – auroral display in 1859, scientists began to suspect a link between activity on the Sun and the appearance of the northern lights. Spectroscopy emerged and the auroral light was measured. Scientists saw that the light was a spectrum of individual lines of colour, which meant that the light must be given off by gases, not by solid or liquid particles on fire. By the end of the nineteenth century, despite these new ideas and the increasing amount of data available from observations of the aurora, there was still no clear theory as to the mechanism of the aurora.

It was the Norwegian physicist, Kristian Olaf Birkeland, a professor at Christiania (now Oslo) University, who made a breakthrough in auroral understanding. He was also very inventive and made his own experiments to demonstrate his theories as well as commercial inventions to fund his personal interest in aurora studies. He is possibly better known for his invention of a nitrogen-fixing process for fertiliser using electric arcs, though his part in the scientific history of the aurora has become better recognised in recent years.

In 1899, while in his early thirties, Birkeland embarked upon a dedicated expedition to northern Norway to study the aurora. This was his second such expedition; the first, in 1897, had ended abruptly when the team was caught in a blizzard on the way up to a hut on the Beskades mountain. They managed to descend to safety but one of the team suffered badly frostbitten hands, ruining his chances of becoming the surgeon he was in training to be. Two years later, Birkeland returned to the area with three others to spend the winter. Two small observatory buildings had been erected on nearby peaks in the Haldde mountains, across the fjord from Alta, during the summer. The separate locations were necessary to perform triangulation and measure the height of the aurora. Magnetic measurements made during the expedition led Birkeland to surmise that the aurora was due to electromagnetic effects and streams of particles coming from the Sun. Cathode rays – high-speed electrons – and X-rays had been discovered in the 1890s and shown to be deflected by magnetic fields. Birkeland also saw a link with sunspots.

The 1899 expedition was not enough to develop and prove Birkeland's theories, so he returned a third time to Finnmark in 1902. This time, the observatory there was one of four sites in different countries – northern Norway, Iceland, Spitsbergen (Svalbard), and Novaya Zemlya in Russia. Birkeland wanted to get magnetic measurements from around the auroral oval at synchronised times to allow for global comparison. From this he developed a theory about currents flowing along the magnetic field lines during an aurora, the importance of which was only realised in the 1970s. His expeditions and theories on the aurora were documented in his book *The Norwegian Aurora Polaris Expedition 1902–1903*, published in 1908. It was through these expeditions that Birkeland was able to deduce that the aurora was ultimately powered by the Sun, thus establishing it as an 'otherworldly' phenomenon in a literal sense after all.

Back in Christiania, Birkeland used his new theories to design a laboratory experiment to corroborate his ideas. This was his terrella, a small model of Earth. In an evacuated glass box he put a metal sphere, which he magnetised using a wound coil electromagnet to have a field like that of the Earth. The outside of the sphere was brass painted with a phosphorescent substance that would glow when hit by electrons. Into this 'space' he fired cathode rays and watched as rings of artificial aurora lit up the poles of the terrella. He was able to re-create aurora effects in the laboratory, and this gave him confidence that his theories were correct. However, Birkeland's work was not readily accepted, particularly in Britain. Scientists thought his ideas were a step too far, too unorthodox. It was only decades later that it became clear that Birkeland was fundamentally correct. In the 1960s, satellites observed particles in 'empty' space just as Birkeland had predicted – stellar winds and magnetic disturbances around the polar regions. His work was also a breakthrough in terms of our understanding of the solar system, applying electromagnetic theory outside the realms of Earth. Norwegians are reminded of this every time they open their wallet – Birkeland's face adorns the 200-kroner note, alongside a small drawing of his terrella experiment.

With the advent of the space age, we have equipped ourselves with the key to decipher this fascinating phenomenon, building on the ground-based observations and theories of scientists such as Birkeland to understand further the mechanisms of the aurora. We have moved from myth to the possibility of knowledge.

When I arrived in Alta and met Cecilie she took me to stay with her aunt. The long windows of her sitting room looked out over the fjord and to the mountains beyond. As I stood looking out at the view, with the light fading over the waters, Cecilie said to me, 'that's Haldde over there,' pointing across and slightly left; 'there's an old observatory up on that mountain. You can just about see it.' There it was right in front of me, the very mountain where Birkeland had over-wintered and made those pioneering measurements to develop his ideas on the northern lights.

I left northern Norway and flew to Oslo, where I met a group for winter training. I had decided to undertake a short ski trip across

Spitsbergen the following year. I wanted to get a taste of what the Arctic explorers of old would have endured and I was hoping to catch a glimpse of the northern lights in the open wilderness.

We drove for a couple of hours up to a cabin in the mountains, our starting point for five days spent learning techniques for camping in snow and skiing with pulks (low sleds, without runners, that are dragged over the snow). The guide, Svante* Strand, entertained us with stories of skiing past polar bears on Spitsbergen and showing us photographs of journeys to the North and South Poles. We had met before, the previous year on Mount Elbrus, the highest mountain in Europe.

The daytime was for skiing. The weather was clear and blue, and, in the prepared ski tracks near the cabin, I took my first tentative slides on narrow Norwegian mountain skis. These are similar to cross-country skis in that the boots are only attached to the ski at the toe, leaving the heel free to lift as the skier slides the back ski forward and through in a stepping motion. The skis themselves are about 5 or 6 centimetres (just over 2 inches) in width, so wider than cross-country skis but narrower and straighter than the alpine (downhill) skis with which I was most familiar. Being a skier already, I was well balanced and comfortable on the skis when moving on the flat or uphill, but I fell a few times on the downhill sections, especially when I tried to turn. As a regular downhill skier it was a strange feeling; the terrain was easy, and yet on these narrow, straight skis with my heel flapping free and my feet moving in my soft boots I had so little control. We amused ourselves skiing in short trips from the cabin and practising up and down the short hills, before heading off for a three-day camping trip in the environs, never too far from the cabin.

The days flew by in a blur of snow, red tents and stories shared amongst the group whilst skiing. In the evenings back in the cabin we sat and talked of mountains and expeditions, watching the

* The 'e' in Svante is pronounced like 'a', so 'Svanta'.

flames in the woodburner as they greedily consumed the wood we fed it. I knew it would be hard to leave. I was falling in love with the north. All of it.

It was not just the explorers who experienced the aurora in harsh conditions. The early auroral researchers braved the snow, ice and wind to make painstaking records of the shapes, colours and movement of the northern lights. Around the turn of the twentieth century, at the same time that Nansen, Amundsen and Scott were taking up skis and exploring new lands in pursuit of the poles, scientists looking skywards in the Arctic regions were beginning to unravel the science of the aurora, casting aside the old stories that had dominated for centuries. I wondered then, thinking about this new, pioneering activity, why it is that mythological explanations of the northern lights are often viewed as more compelling than the scientific explanations. It struck me as odd that the idea of charged particles flung out and journeying for two days from the Sun, being accelerated by the Earth's magnetic field and causing massive disruptions in our upper atmosphere, could be considered more prosaic - require less imagination - than, for example, a magic fire-fox! The real story of the aurora is complex and technical, but no less wondrous or fantastical. The creativity and originality of thought required by all the scientists who contributed to the understanding of the aurora is immense. This is only the beginning.

CHAPTER THREE

ICELAND – ROCK, ICE AND FIRE

I WOKE UP AND pushed my sleep mask away from my eyes. Despite the mask, my eyes still felt gritty, and after I wiped them I looked down at my hands and saw my fingers black with dust. The wind was still whistling outside and my tent was shuddering violently. My tent was tiny – a little one-man, only as wide as my sleeping mat. My rucksack and meagre provisions were stacked at the bottom, by my feet. And everything, absolutely everything, in the tent was covered in a thick layer of fine black volcanic dust. Even as I sat, more was being blown in through the mesh of the tent. It had been streaming in all night, disturbing my sleep as it landed on my face and caked around my nostrils.

I wriggled into my trousers, pulled on a gritty jacket and unzipped the inner tent door. The outer part had escaped from the tent pegs and was flapping in the wind. I pulled on my boots and emerged. Outside, people were in various stages of packing up tents, struggling to control the light fabrics in the wind. It was too windy to cook; plus I had eaten enough dust in last night's dinner. I decided to pack up and leave as quickly as possible.

I went back into my tent and hurriedly wiped sand off my exposed belongings and stuffed them into my rucksack. I fought with my inflatable mat in the confined space of the tent, trying to deflate it enough to squeeze it into its sack in the shortest time possible. Then I stood up, achy and stiff from yesterday's walk, and tried to shake as much dust from the tent as the wind would allow. The flesh around my hips was bruised and sore from carrying the heavy rucksack. This, I thought, had to be the low point of the whole hike. I understood now why the huts were booked up so far in advance: camping in a volcanic landscape had its own unique challenges. As I bent down to remove the remaining tent pegs, my tent pitched violently from side to side in another big gust of wind and inhaled as much dust again as I had just shaken out.

I was camping at Botnar in the desolate highland pasture area of Emstrur, southeast Iceland. It lies northwest of the large – but not the largest – icecap Myrdalsjökull, neighbour to the smaller, but infamous, Eyjafjallajökull*. I was three days into the four-day Laugavegur trail hike from Landmannalaugar to Thórsmörk. Except I, along with four companions, had just walked here in two days, which was the reason we were so bruised and tired. I had come to Iceland to see the unique geology of this small volcanic island and what it could tell us about the Earth's magnetism, which is a fundamental component in the generation of the northern and southern lights. Only planets with magnetic fields will ever exhibit aurorae. I had also come to Iceland to learn about volcanoes and to think about the four different states of matter, of which fire and ice are the extremes. Like fire, the Sun is a plasma, the fourth state of matter, as is the aurora. It is only by interrogating its properties that we can glean deeper insight into the capricious nature of the polar lights.

The trail began in Landmannalaugar, a busy encampment in a wide, flat valley four hours' bus ride east of Reykjavík. The flat valley floor was surrounded by hills that looked like giant piles of builders' sand, the orange-brown earth streaked with light and dark watermarks and the more protected areas sporting a fine covering of fluorescent green moss.

---

* Pronounced (approximately) ay-ya fee-ay-ya yoh-kult, if I am remembering correctly from lengthy instruction by a friendly waitress in a Reykjavík bar.

A boardwalk led out of the camp over a steaming-hot stream where walkers wallowed, whilst a sign pointed along the Laugavegur trail towards Hrafntinnusker. I set my boots to the trail. Over the following three days I would walk 55 kilometres (34.2 miles) through a varied landscape; from active, steaming, desolate hills south to the grasses and trees of the protected, vegetated valley of Thórsmörk. We began traversing mountains of orange rhyolite and fields of glossy, sharp obsidian stacked like rubble, gaps filled in with fine, dark sand and growing with the ubiquitous fluorescent moss, the brightest colour in the drizzly muted landscape until the sun came out and the rhyolite shone. We walked past steaming vents, where the surrounding ground was caked with thick, dirty-white sulphur, and through steaming, misty patches where smaller, distributed vents exhaled their warm breath gently onto the breeze. We crossed areas of lumpy basalt mounds like one-piece cairns, adorned with mosses and lichens. The near-constant drizzle on the first day kept the fine sand of the path under control. In several places hot streams bubbled and hissed from small openings in the rock and steam rose from the earth, expanding into cloudy plumes before dissipating into fine mist. Patches of snow displayed the wave-like patterns of the wind, highlighted by the black dust on their peaks. Monochrome contouring everywhere. From here, we dropped down into the real blackness, the dark fields of basalt and dust. These stretched on for about 10 kilometres (6 miles), a flat floor of ancient lava surrounded by the streaky hills and riddled with slate-grey rivers, fast-flowing and only traversable by wading. It was water so cold that the chill penetrated our feet instantly, causing sharp pain and involuntary squealing as walkers pushed through the flow, which sometimes came to above the knees. It was from this seemingly interminable lava field that I descended into the dustbowl of the campsite. As I packed up my tent I was keen to reach the more protected, grassy – and less dusty – region of Thórsmörk.

The colours of Landmannalaugar come from the rhyolite rocks – relatively rare igneous, silica-rich rocks produced in volcanoes in different forms depending on the eruption. A pale-coloured, foam-like pumice is produced in explosive eruptions where lots of gas bubbles get into the lava. It may produce a grey-yellow, very hard rock, or even the black, glassy obsidian if the lava cools too quickly for crystals to form. But that is just the beginning of the process. In the ground over a long, long time, when subjected to heat and water percolation, the rhyolite can change to form brown and red, even pale greens. The Landmannalaugar area is Iceland's largest hot geothermal field and is dominated by rhyolite that has been around for half a million years. Along with the abundance of water coming through towards the surface, the geothermal energy alters the rocks over time, changing the minerals – and colour – within it.

The most colourful rhyolite is relatively old, but the obsidian sitting on top is young. Around Landmannalaugar the obsidian came from an eruption at Torfajökull in 1477. There was also an obsidian field just before the huts at Hrafntinnusker, at the end of the first day, this one dating back to eight or nine hundred years ago. The shine of obsidian meant it was coveted for ornamental purposes, and it is still a semi-precious stone today. The obsidian around Hrafntinnusker is some of the purest in Iceland and it gives its name to the vicinity, the word meaning 'obsidian peak'.

Iceland is relatively young in Earthly terms, and geological activity is changing its landscape more rapidly than in other places on the planet. The landmass formed about 25 million years ago. The Earth itself grew from the coalescence of solid material in a rotating, dusty disc around the newly formed Sun over 4.5 billion years ago. As the Earth was growing from a marble-sized rock into a billion-trillion-tonne planet, it went through collision after collision with meteorites and other planetesimals. Such impact generated heat and rocks melted, then this fluid, molten rock separated out according to density, just as oil and vinegar separate out in salad dressing. Heavy iron fell to the centre, forming the Earth's core – solid at the very centre and liquid further out. Less dense silicate material floated on top, creating the mantle. Finally, the silicate portion separated and a very thin, solid, rocky crust formed at the surface.

The crust floats on the denser, flexible rock of the mantle like an iceberg floats on water, and the crust varies in depth in the same way. Taller mountain ranges need thicker crust to support them. Convection currents in the mantle, essentially heat rising from the centre of the planet, drive the movement of rigid sections of the Earth's crust (and a tiny slice of the uppermost mantle) called plates, and the process of this movement is known as plate tectonics. Plate tectonics can cause plates to collide with each other, from which mountain ranges are created or volcanism ensues. Oceanic crust can also be subducted, or pushed underneath another plate, so that it is recycled into the mantle. Melting of the rock deep in the Earth triggers volcanoes and earthquakes as the molten rock rises up through the crust, exemplified in the 'ring of fire' that surrounds the Pacific Plate, which is home to 75 per cent of the world's active and dormant volcanoes. Where two pieces of deeply rooted continental crust collide, neither can pass beneath the other, so the rock crumples and buckles, forming vast mountain ranges like the Alps or the Himalayas. Conversely, where plates diverge, the space is filled with magma that wells up from below and solidifies to form crust. Such areas are typically mid-ocean ridges. This sea floor spreading means that the continental crust is much older than the oceanic crust because the oceanic crust is continually being renewed.

Iceland is unusual because it lies on one of these spreading centres, the mid-Atlantic ridge, and it is the only place in the world where an oceanic ridge is seen coming up out of the ocean. It also sits on a hotspot, where hot magma from the mantle rises up in narrow columns in a mantle plume. In Iceland, this plume currently rises in the region of the biggest icecap, Vatnajökull. Around 25 million years ago, the mantle plume drove the mid-Atlantic ridge up to the ocean's surface and brought up extra heat to melt crust rock and create the island. To put Iceland's relative youth into perspective, if the Earth were only one year old then Iceland would have been born just two days ago. In that scheme of things, the Ice Age glaciers appeared about five hours ago and only melted away in the Holocene a single minute ago.

Iceland's rate of geological activity means that most of the rocks are even younger than 25 million years old. In fact, the oldest rocks on the surface date back 16 million years and sit at the peripheries of the island, away from the mid-Atlantic ridge. Rocks nearer the ridge are much younger, and the rhyolite around Landmannalaugar dates back only about half a million years. Most of the rock in Iceland is basalt. In fact, most of the rock on Earth is basalt – the common, dark, volcanic igneous rock that comes up at spreading centres. All volcanic islands are made from it, and Iceland is no exception.

The area around Landmannalaugar through which I walked contains some of the largest rhyolite deposits in Iceland and is a prime crust-processing location. Its unusual geology is due to a fault in the rift zone. The mid-Atlantic ridge passing through Iceland is not straight; it comes into Iceland in the south-west near Reykjavík and exits to the north of Iceland just east of Akureyri. This fault encourages deep, older crust to be dragged south and allows rock melt to rise up, forming the active region of the Torfajökull icecap, just south of Landmannalaugar, where rhyolite eruptions create the beautiful and various colours for which the area is best known.

The volcanoes and icecaps of Iceland are there 'because something very interesting is going on at depth,' says Dave McGarvie of the Open University in the UK. Activity deep within the Earth is responsible for Iceland's unique geological history. It is also the cause of the Earth's magnetism, or geomagnetism, without which the aurora on Earth would not occur.

Humans have been familiar with the basic concept of Earth's magnetism since ancient times, but it was its use in navigation and exploration that piqued the European interest and led to a better understanding of the phenomenon and its complexities. The humble compass is one of the simplest, and earliest, magnetic instruments there is, and it was invented in ancient China. A freely suspended magnetic needle gently spins so it is aligned north–south, the end that we have labelled north-seeking (and generally painted red) pointing to geographical north. However, navigators in the fifteenth century realised that the compass doesn't point north everywhere; there is some magnetic variation, both spatially around the globe and over time. This magnetic variation is greatest at the poles.

We didn't find out why this geomagnetism exists at all until much later, once the field of seismology had emerged in the early twentieth century. Seismology is the study of the waves produced by earthquakes, called seismic waves, as they travel through the Earth. Speculation about earthquakes goes back to ancient history, but there was no systematic study until a large earthquake occurred in Europe in 1755, after which catalogues of historical events and related observations were recorded. In the mid- to late- 1800s, seismology emerged as a separate discipline and scientists began measuring earthquake waves using seismographs and developing theories on the behaviour of such waves. By the early twentieth century, understanding had increased sufficiently that the waves could be used to map the interior of the Earth. From the differing propagation of the two varieties of seismic waves (P and S), seismologists were able to infer that the Earth has a fluid interior, and from this they were able to deduce the layered structure of the Earth as described previously: a solid inner and a fluid outer core; a viscous mantle; and a solid crust. This important breakthrough had implications outside the field because the movement possible in a liquid layer allowed the formulation of a dynamo theory to explain the Earth's magnetism.

Studies of rock density, geomagnetism and meteorites have concluded that the Earth's core is composed mainly of iron with some nickel, and perhaps some sulphur. It is the movement of the iron in the liquid core that generates the geomagnetic field. Heat in the Earth's inner core generates convection currents in the outer core. As fluid is heated, its thermal buoyancy changes and it becomes unstable. Individual parcels of fluid will absorb heat, become buoyant and float to the top, where they will liberate their heat and sink to the bottom again. So the cycle continues, creating the motion of convection.

Iron, as a metal, conducts electricity. It does this by means of so-called free electrons that are disassociated from individual atoms and can move about freely within the metal. The movement of this electrically charged molten metal in the outer core of the Earth constitutes an electrical current, and this electrical current has an associated magnetic field – the geomagnetic field.

The magnetic field generated is dipolar, so it looks like that of a bar magnet, which has a symmetric field pattern that looks like butterfly wings around its bar-magnet body. This comes about because of the flow pattern of the molten iron in the outer core. Convection sends the fluid outwards and back in loops from the central core, whilst the rotation of the Earth twists it west–east. This results in the fluid flowing fastest in spiral channels aligned approximately north–south, these spirals acting like huge coils of electrical wire and creating a magnetic field like a big bar magnet. This is what the geomagnetic field looks like, and it extends up out of the core, right through the mantle and up to the Earth's surface and beyond into space.

The motion of a liquid is often turbulent and unpredictable, and the outer core is no exception, which means that the geomagnetic field slowly drifts and changes over time. This was noticed by early cartographers, often colonial Europeans, mapping the Earth and the magnetic field. Moreover, the geomagnetic field sometimes reverses polarity. In other words, if you were standing with a compass facing north as the flip happened, you would see the red, north-seeking end of the compass needle (that was originally pointing directly ahead) rotate to face south, behind you. This magnetic flip over geologic time was first noticed by Bernard

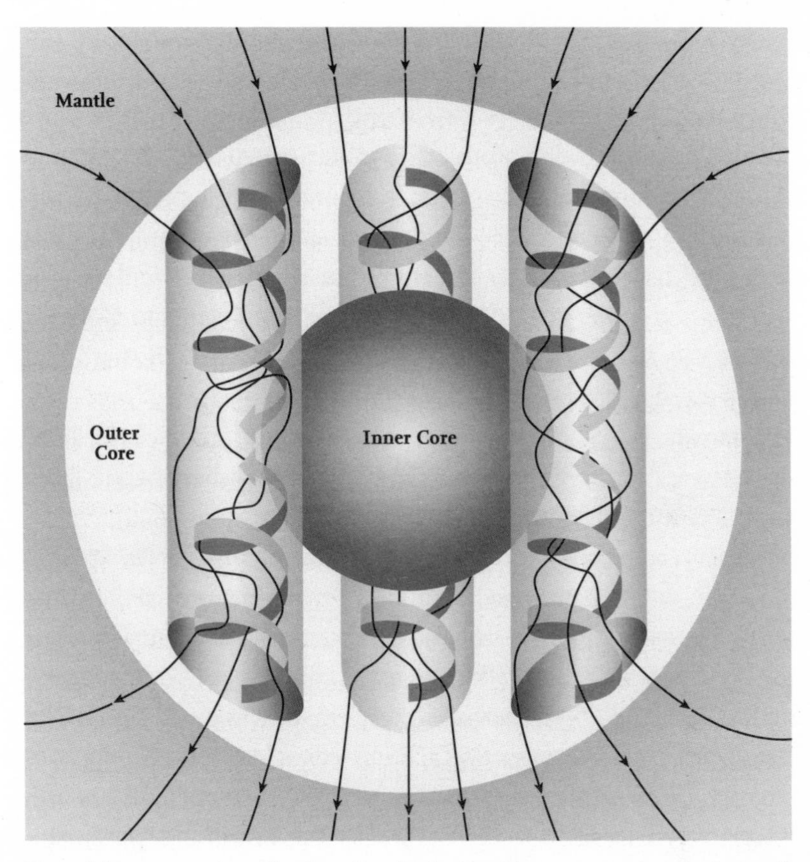

Heat-driven convection of molten iron in the Earth's outer core is spun into rolls by the Coriolis force. These circulating electric currents generate the Earth's magnetic field.

Brunhes in France in 1904. He was measuring magnetisation in basaltic lava flow and noticed that some of the rocks had fields aligned in the opposite direction to those of the Earth. As molten rock cools, susceptible particles within it line up along the Earth's magnetic field, preserving an imprint of that particular field alignment forever. But back then, the geomagnetic field was not well understood and the magnetisations of rocks being found were difficult to correlate when they came from volcanoes separated by great distances and from sometimes indistinguishable time periods.

By the early 1960s, radiometric dating had improved significantly from its origin in 1905, particularly with improvements in mass spectrometry during the Second World War, which enabled scientists to determine atomic and molecular masses. Also called radioactive dating, the technique measures the proportion of radioactive isotopes and their products in a sample and compares this with known decay rates to determine the age of the sample. Using this method in the 1960s, Allan Cox, Richard Doell and G. Brent Dalrymple, scientists at the United States Geological Survey, had begun dating magnetised rocks to build up a timeline of reversals. In this way, by looking at the geological record, they could build up a history of the Earth's magnetic field. What they found from such a picture is that the magnetisation had reversed many times, but not in a regular, periodic way. Some reversals were of long duration, generally around a million years or more, which they termed epochs; others were shorter, about a tenth of the length of the epochs, which they called events. There may be short reversal events within a longer epoch, each of which was given a name by Cox, Doell and Dalrymple that acknowledged an eminent geophysicist who had contributed to the field. We are currently in the Brunhes epoch, and the last reversal – the Brunhes-Matuyama – occurred 780,000 years ago.

The geodynamo is a chaotic system, so it is inherently unpredictable. We don't know when the polarity will reverse again. However, in the mid-1990s Gary Glatzmaier and colleagues from the US Institute of Geophysics and Planetary Physics created a supercomputer model of the geodynamo that mimics the real Earth. From this they learned that the magnetic field can take five thousand years or so to make the transition from one polarity to another. It's interesting to think about what might happen to the aurora during such a reversal, since we know that charged particles are channelled down magnetic field lines. It's likely that the northern lights will no longer be in the north, and the southern lights no longer in the south. During the reversal, the geomagnetic field would not vanish entirely; the north-south-oriented dipole field would disappear, but, to compensate for this, other more complicated parts of the field would grow stronger. New magnetic poles could pop up anywhere, bringing the aurora to a wholly new part of the world.

At around the same time that Allan Cox became interested in field reversals, in the 1950s and 60s, magnetic data was collected by ocean-going research vessels. When plotted on a map, scientists saw that the ocean floor was a collection of magnetic stripes of alternating polarity, looking a bit like a symmetrical geological barcode. And they are centred on mid-ocean ridges. The stripes fitted perfectly with the theory of sea floor spreading that had been recently proposed by Harry Hess and by recent work by Bodvarsson and Walker in 1964, who had been studying new crust forming at the Mid-Atlantic Ridge in Iceland. Magma wells up from the mantle, pushing the crustal plates apart and encoding the Earth's magnetic signature as it cools into basalt rock. Matching these up with Cox's timescale, Fred Vine showed that the sea floor was spreading at a constant rate. It was this evidence that confirmed the ideas of plate tectonics described earlier as credible theory, rather than mere speculation. Icelandic rocks and geology played a major part in the development and proof of geomagnetism and plate tectonic theories, because Iceland's relative youth and still-active situation meant that there are hundreds of thousands of fresh basalt lava flows accessible for sampling. Additionally, Iceland, sitting astride the Mid-Atlantic Ridge, allowed scientists to envisage the mid-ocean ridges inaccessible under the sea.

At the end of the Laugavegur trail in Thórsmörk, I was in the shadow of Eyjafjallajökull, the stratovolcano that erupted in 2010, causing global flight chaos and the largest air-traffic shutdown since the Second World War. The eruption started beneath the icecap in the early morning of 14 April. A small flood of meltwater occurred on the south side of the volcano and then, over the course of the day, the pressure rose under the ice until in the evening there was an explosive release of magma, amplified by the interaction with the ice, that created a plume stretching almost 10 kilometres (6 miles) into the sky. Winds in the upper atmosphere carried the ash-rich cloud towards Europe and America, easily dispersing over a wide area because the ash was particularly fine and light. The problem for aircraft was that the ash included a mixture of coarse and fine glass-like particles that acted like a sandblaster on jet engines. Millions of people were stranded at airports as aeroplanes were unable to fly, though the Icelanders themselves were comparatively unaffected. People in the vicinity were evacuated during the explosive phase and a layer of ash up to 10 centimetres (4 inches) thick in places settled over everything, but the experienced Icelanders took it all in their stride. The eruption lasted six weeks, and a huge clean-up was required afterwards, the fine dust having seeped into everything. I could empathise, having spent that uncomfortable night in the volcanic dustbowl of the Botnar campsite.

The Eyjafjallajökull eruption caused havoc on a global scale and serves as a recent reminder of the power of nature. We may do our utmost to understand, predict and mitigate these events and their effects, but we are not in control. We bow to higher powers, both within our own Earth system and outside.

We are also reminded that the Earth is not static and is instead in constant, though sometimes slow, flux. Iceland is one of the more obvious examples of our dynamic planet – it is still in formation and likely to remain active for millennia to come. It has around 30 currently active volcano systems, and from the time of the Viking settlement to the present day there has been a volcanic eruption on average every five years. Its northerly location means the majority of volcanoes are covered in glaciers or icecaps, which produce the explosive eruptions like that of Eyjafjallajökull. As I walked the last kilometres of the Laugavegur trail, past ice and snow patches blackened with volcanic dust, I thought that Iceland really is the land of ice and fire.

* * *

ICE AND FIRE lie at the two extremes of the states of matter – solid and plasma. On Earth, we live in a tiny pocket of the Universe dominated by the first three states: solids, liquids and gases. But the majority of the Universe actually exists in the plasma state. Over 99.99 per cent of the known, visible Universe is plasma – also known as charged gas, or the fourth state of matter.

Of course, everyone is familiar with solids, liquids and gases, and in fact you are probably more familiar with plasma than you might realise. The Sun is made up of plasma; plasma forms lightning. Flames and neon lights are plasmas that people are in contact with almost every day. Perhaps you have a plasma television at home. The aurora is also a plasma – a diffuse, low-density one that appears in the ionised upper part of our atmosphere (which, incidentally, is itself a plasma). In order to understand the aurora more fully, we must first get to grips with this fourth state of matter.

Let's start with the familiar states. Substances change state – from solid to liquid, liquid to gas and ultimately gas to plasma – by the application of heat or energy. On a molecular or atomic level, this heat increases the movement of the molecules that make up the substance. In a solid the molecules are all in fairly fixed positions, held together by intermolecular forces. The molecules have *some* energy, so they vibrate a bit, but their positions are fixed and so the substance has a determined structure; it is solid. Heat this solid, though, and the molecules start to vibrate faster and faster as they get more and more energy. Ultimately they break away from their fixed positions and the substance can flow; it is now a liquid. The molecules still experience the intermolecular forces that try to keep the substance together, like the surface tension that causes water to bead on a surface, but the particles are moving so much faster now that those forces have less effect. The more the temperature rises, the faster the molecules move.

As the liquid continues to be heated, the molecules move faster and faster, until some have enough energy to escape from the surface of the liquid, becoming a gas. Now the particles no longer experience the intermolecular forces at all so they spread out unconstrained, filling whatever space they have available to them.

So far, so simple. This is what happens in the first three states of matter. We could call them the Earthly states, though that's not to say they don't exist elsewhere in the Universe. There are other planets, molecular clouds, icy dust grains and other parts of space where the first three states can be encountered, but our own experience of them is Earthly. Plasma, the more exotic fourth state, even when encountered on Earth, has a more ethereal quality.

As we have seen, the more energy (or heat) a substance receives, the faster its particles will move. The speed of particle movement is a measure of the temperature. Daniel Bernoulli first propounded this idea in his 1738 work *Hydrodynamica*, which argued that gases consist of a vast number of molecules moving in a random motion. The impact of these molecules on a surface, such as our skin, is what we feel as pressure, and their kinetic – or movement – energy is what we feel as heat. Bernoulli demonstrated that pressure and particle motion increase with temperature, thereby setting down the foundations for the kinetic theory of gases.

Over a hundred years later, in which time the theory had developed a bit, James Clerk Maxwell developed a statistical representation of the molecules' behaviour. Published in 1860, this gave a probability description to the motion of the particles, a means of calculating what proportion of molecules would have a velocity within a certain range. Maxwell realised that a molecule may have any speed, but that some speeds are more likely than others. Their speeds are distributed in a lopsided bell curve*, with

* A bell curve is the informal name for a graph where the value of some measured quantity, like speed, increases along the bottom of the graph and the probability of that value occurring, indicated by height, increases to a peak and then drops away again, tracing out a shape similar to a bell. A standard, symmetrical bell curve is called a Gaussian or normal distribution. The Maxwell-Bolztmann distribution for particle speeds is similar in shape but not completely symmetrical.

the majority having a medium amount of energy relative to the rest. Smaller and smaller numbers of particles will have extremely low or extremely high energies, with these particle numbers making up the tapering edges of the bell. He realised that there are always slightly more fast-moving particles than slow-moving ones. This is because the lowest possible speed is zero whilst the highest speed is not constrained at all (except by the speed of light – the 'speed limit' of the Universe), so this means there will always be slightly more faster particles than slower ones, just because there can be.

The theory shows that as the temperature of the gas as a whole increases, the bell flattens and the peak of the bell moves right, indicating that the most probable speed of particles in the gas is now higher and that the average kinetic energy of the gas is also higher. In other words, the temperature of a gas is proportional to the average kinetic energy of the particles; it's therefore related to the square of the particle speeds and their masses. Ludwig Boltzmann formalised Maxwell's idea of distributed speeds in 1871, and the bell curve of particle speeds became known as the Maxwell-Boltzmann distribution. So, in general, giving a substance more energy increases the speed of movement of its constituent particles, allowing them to overcome intermolecular forces and change state. How quickly they can change state also depends on particle mass and the consequent strength of the intermolecular forces.

Plasma is special. It differs from the other states in that extra energy does not cause the particles to move faster; it actually *breaks them apart*. The atom is stripped down to its core, with the extra energy overcoming the electrostatic forces that bind the elements of the atom together, removing the electrons one by one. In a neutral atom, the positively charged protons in the nucleus exactly balance the negatively charged electrons that flit immeasurably around in the electron cloud. Each time an electron is stripped away, the remnant atom is left a little more positive. This process of electron-stripping is called ionisation, and the remnant atom is called an ion. With sufficient energy the atom can be fully ionised, with all of its electrons having been stripped away to leave just the nucleus of the atom. In this way the original gas becomes a 'soup' of ionised particles – positives and negatives all swarming around independently. A plasma is therefore a *charged gas*, and it has some very interesting properties by virtue of these free charges buzzing about within it.

In a plasma every charged particle pulls or pushes on every other charged particle, so the actions of individual particles affect each other. This is known as *collective behaviour* and is one of the major characteristics of a plasma; one that separates it from the other three states. Think about a plasma globe and how the flickering tendrils of purplish light that move between its centre and the outer glass will follow a finger or a hand that is held alongside. If the glass globe was filled with a solid such as ice, or a liquid such as water, or even a gas such as steam, would putting your hand to the glass have any visible effect on the matter inside at all? No. Yet with a plasma inside it does, because each charged particle is pushing or pulling on every other charged particle, so as the charges nearest the glass respond to the change in electric potential created by your finger, they affect the other particles nearby.

Many people first come across the word 'plasma' in a biological or medical context. It is the colourless fluid in blood, in which the red and white blood cells (or corpuscles) are suspended and move. In fact, it is precisely this plasma from which electromagnetic plasma gets its name. Irving Langmuir, an American physical chemist, coined the term when experimenting with ionised gases in the late 1920s. The coordinated behaviour displayed by the electrons and ions, dancing together in a clear formation while the underlying music was yet to be heard, reminded Langmuir of blood, where the corpuscles are carried in the fluid plasma. Analogously, he called this new and unusual substance 'plasma', feeling that something thus far unseen must be responsible for the particles' behaviour. In time, scientists came to learn more about the dance, and how the intricate movement was due not to an underlying fluid but to electromagnetic forces acting on and between the charged particles themselves.

Collective behaviour in plasma causes rippling waves to travel through the medium, as the free charges move to even out any areas of charge imbalance. Despite these free charges, a plasma in general is almost neutral (or *quasineutral*), because opposite charges attract. When too many of one charge accumulate in an area, electrical forces are set up that pull opposing charges back in so that neutrality is regained, meaning that the plasma as a whole remains broadly neutral. However, as these mobile charges speed into a new area they overshoot their destination and are pulled back again, like a pendulum swinging perpetually past its resting point, causing waves in the plasma – small, fast, rippling waves of the tiny light electrons and large, lolloping waves of the heavy ions.

This collective behaviour means that plasmas can be quite unstable. Little disturbances that would be insignificant individually can be passed along and amplified in a collective situation. Imagine looking from a distance at a million people walking along a vast, exposed mountain ridge with sheer drops at either side. If one or two people fell off we might not notice – they are a mere fluctuation in such a large group of people. But imagine if those million people were all roped together on the mountain. Now if one or two were to fall off they would pull others with them, causing a large and serious crisis of tumbling bodies on the mountain. In a similar way, small disturbances in plasmas can quickly develop into large, dangerous instabilities.

Plasmas are dominated by electromagnetic forces. It's an effect of the charged particles, again, and of the interplay between electricity and magnetism. It was the English physicist Michael Faraday, the self-taught son of a blacksmith, who first rigorously investigated this interaction. He was already aware that electric currents could produce magnetic effects. For example, an electric current in a wire would deflect a magnetic compass needle. Faraday wanted to show that the opposite effect was also possible. Throughout the 1820s he worked on producing an electric current by magnetic action, and in the early 1830s he succeeded in doing so, showing that electricity and magnetism do indeed interact.

Moving charges generate magnetic fields around them, so an electric current (which is a flow of negatively charged electrons) generates a magnetic field. The French physicist and mathematician André-Marie Ampère demonstrated this in 1825. He showed that two wires carrying electric current flowing in the same direction would repel each other in the same way that two north poles (or indeed two south poles) of a magnet would. Similarly, when the current flows in opposite directions in the two wires they attract each other. This is now known as Ampère's law after he published the mathematics of the effect in 1827. It is also the principle of the electromagnet. A magnetic field is generated around a wire, so if this wire is coiled into a long spring-type shape (known as a solenoid) then the magnetic field will be intensified with every turn. The magnetic field passes through the centre of the solenoid and loops round to the other side, forming the old, familiar butterfly-wings field pattern of a bar magnet. Wrapping the wire around an iron core further increases the magnetic field by magnetising the iron, too, with the individual molecules in the metal all aligning with the magnetic field.

Faraday used this principle to make a generator. He caused a current to flow simply by moving a magnet. His apparatus consisted of two solenoids situated close to one another, one connected to a battery and the other to a galvanometer, which measured the flow of the current. When he connected and disconnected the battery, so causing current flow to change in the first circuit, a current registered in the second. This current in the second coil was entirely caused by the magnetic field created by the current flowing through the first. This effect is now known as electromagnetic induction, and it is used in many useful yet seemingly mundane applications such as in electric motors, on rollercoaster braking and for powering contactless cards.

To conceptualise the ideas of electromagnetism, Faraday introduced his own construct, known as lines of force, which together constitute a 'field'. The familiar butterfly-wings pattern of magnetic field lines is an expression of the idea invented by Faraday, showing the direction and the strength of the magnetic effect. He saw the interaction between electric and magnetic effects as if the lines of electric force intersected those of the magnetism, so vibrations along one of the lines could be communicated to the other.

Faraday was a great experimental scientist but he wasn't a mathematician. However, his theories were adopted and advanced by others more mathematically inclined. Maxwell, who had previously developed the statistical ideas of gas dynamics, unified Faraday's theories on electromagnetic induction with Ampère's law and Gauss's laws to produce a concise theoretical description of the fundamentals of electricity and magnetism, one that was based entirely on experimental results. Maxwell's equations are among the most important equations in plasma physics, and they have an interesting conclusion. They describe what gives rise to electric and magnetic fields, and what configurations those fields can take. Since changing magnetic fields produce electric fields, and changing electric fields produce magnetic fields, Maxwell demonstrated that oscillating electric and magnetic fields would produce waves that would propagate throughout the medium. He calculated the speed of these waves, and found it was very close to the experimentally measured speed of light. In effect, Maxwell had found that visible light was an *electromagnetic wave*, an oscillation of electrical and magnetic fields driving each other perpetually as they travel through space. The speed of these waves, he showed, would be the same regardless of the rate at which the fields were changing, and no matter how fast the source of the waves itself was moving. It was this finding that inspired Einstein's Theory of Special Relativity.

Plasmas are dominated by electromagnetic effects because they are composed of freely moving electric charges. The moving charged particles generate magnetic fields and, conversely, magnetic fields also affect the charged particles, imparting a deflecting force that curves them off their straight path and sends them spinning in helical twists around their direction of travel. All plasma particles travel in spirals. The radii of the circles that the particles trace are proportional to their velocity and their mass. In other words, the hefty ions trace out much larger orbits than the nippy electrons. In a plasma, these electromagnetic effects are constantly in action, feeding back on each other to modify the particle movement and collective plasma behaviour, creating a dynamic entity.

Plasmas conduct electricity by virtue of the free charges they contain. This is a very special property, and is one of the features that differentiates plasma from an insulating gas. In order to conduct, the plasma does not have to be fully ionised; just a few ionised atoms are enough to start an escalation of conduction. They excite other nearby atoms, freeing more charges. With just one atom in a thousand ionised, a gas can reach half of its maximum conductivity. Lightning occurs in just such a way. During a storm, convection currents in the atmosphere move water droplets and ice crystals past each other, building up charges in the clouds, in much the same way that a girl rubbing a party balloon on her cardigan builds up static charge that goes on to attract her hair to the balloon when she holds it above her head. The different regions of collected charge set up electric fields, positive to negative, and when the built-up charge and electric field gets too much a tiny current starts to flow, ionising the air molecules and opening up a conductive path. A flash of lightning restores the balance. Peak currents of about 30,000A can flow to Earth with enough energy to toast 100,000 slices of bread, according to a calculation by the UK Institute of Physics. The clap of thunder is caused by intense heating down the lightning channel as temperatures reach tens of thousands of degrees, with the resulting pressure change sending a shockwave rippling through the atmosphere. The unpredictable, intense nature of lightning makes it one of the most dangerous of all plasma experiences.

The common properties of collective behaviour, quasineutrality and the domination of electromagnetic forces unify plasmas over a wide range of scales, energies, temperatures and densities. Plasmas are found in a variety of laboratory and industrial processes, from the very small – such as etching microchips for computers, or in microscale plasma-based transistors used in some high-voltage light sources and medical devices, which are only the width of a few human hairs – to the very large, such as the plasmas that are created in experimental nuclear fusion devices. The Joint European Torus, for example, has a plasma volume of 200 cubic metres (7060 cubic feet), and the ITER ('The Way' in Latin) machine currently under construction will have a volume of more than 800 cubic metres (28,250 cubic feet), about a third of that of an Olympic-sized swimming pool. But these are just the man-made plasmas – natural plasmas exist on staggering scales. Earth-bound ones like lightning and the aurora are, relatively speaking, small. Stars, which are huge balls of plasma, are millions of times bigger than the Earth, and the cold interstellar plasmas between stars can stretch for trillions of kilometres. Yet the theories developed to explain plasmas apply equally on both micro- and mega-scales, so studying plasmas in the laboratory can provide a better understanding of astrophysical behaviour, where it may not be possible to make measurements directly. Unfortunately, plasmas are exceedingly complicated because of their fluid nature, combined with electromagnetic feedbacks, so many approximations have to be made to simplify the theory and the calculations. There are still mysteries and unexplained phenomena that don't fit the theories. One of the biggest of these is the Sun – arguably the most important kind of plasma known to humankind – and its surprisingly hot atmosphere. But the Sun is also pivotal in our story of the aurora. If we want to understand the northern lights, we need to start right back at the Sun.

* * *

THE SUN IS a huge, superheated ball of plasma; a gravitationally confined fusion reactor in the sky. It plays an important part in all of our lives, being the centre of our solar system and the star around which we orbit. The Sun is always in the sight of Earth and we live in its domain: it influences us in profound ways, ways that go beyond the mere giving of day and night. From Earth, viewed askance with the naked eye, the Sun seems little more than a shining, featureless orb, but look closer and you will see it is very far from uniform. This variation affects us on Earth. We have only relatively recently in human history – since the advent of the space age – been able to come to appreciate its dynamism, yet we are still so far from understanding it.

If you have seen pictures of the Sun from images taken by satellites such as the NASA Solar Dynamics Observatory, or the drawings made by Galileo in the early seventeenth century, then you will have seen sunspots. These are dark patches on the surface of the Sun that move across the solar disc as the Sun rotates, disappearing and coming back round again about 27 days later. Mankind has been observing sunspots since ancient times, the first actual recordings going back to China over 2000 years ago, but Western science was relatively slow to take notice. Modern records date back to 1610, when Galileo and others began observing with the newly invented telescope. The research was unwelcome then because it was interpreted as undermining scripture and the entrenched Aristotelian idea that celestial bodies were perfect and unchanging. Galileo was famously tried and sentenced in the 1630s for his heretical work on astronomy, which supported the Copernican view of the Earth moving around the Sun.

The first mention of the possible periodicity of sunspots and the tentative suggestion that they may affect other astronomical bodies that are 'lit up by the Sun' came from the Danish astronomer Christian Horrebow in 1776. In his diary, he determined that sunspots must be periodic, but lamented that there was insufficient effort dedicated to regular historical data to discern 'a precise order of regularity and appearance.' He hoped that more frequent observations would be made to this end. Horrebow himself began a fairly regular study of sunspots in 1738, but his data did not suffice to find a pattern. It was not until 1844 that German amateur astronomer Heinrich Schwabe, who had been observing sunspots for the previous eighteen years, showed a cyclic pattern of about 10 years. His work was extended by Rudolf Wolf, at the Bern Observatory and later at Zurich, who, as well as making daily measurements, attempted to enlarge the sunspot data set by filling in gaps in patchy records from previous decades. When primary observation data was unavailable he used geomagnetic activity measurements as a proxy for sunspot numbers. As such, reconstructed data from pre-1849 is useful as an indication of pattern but is far less reliable than the record post-1849. The measurement Wolf started, which is built up from observing sunspot groups rather than individual spots and which is now known as the International Sunspot Number, provides the longest record of solar activity we have.

Since they are essentially blemishes on the surface of the Sun, sunspots give an idea of how large the Sun really is relative to Earth. Our home planet would get lost in an average-sized sunspot. Indeed, we and the other planets are like minute pebbles in the beach-ball Sun's lake*. Sunspots give an obvious, visual indication of the Sun's variability, but are only one small part of the underlying structure and turbulence of the Sun. It is not simply a uniform ball of plasma; rather, it has layers with different properties, just like the Earth. At the centre is the hot, dense core where fusion reactions are taking place. Around this are outer layers of differing densities where different means of energy transfer dominate, meaning the path of

* If the Sun is the size of a beach ball then the Earth, which is about a million times smaller in volume, would be about the size of a ball bearing. The atmosphere of the Sun extends out into space for trillions of kilometres, which we could estimate as a volume approximately equivalent to Lake Geneva if the Sun is a beach ball.

a light ray from the centre of the Sun is not a simple one. But first, let's consider the fusion reaction that creates the light ray.

Fusion is the joining together of small atomic nuclei to make larger ones. Stellar nucleosynthesis, or fusion in stars, is the mechanism by which all the heavier elements in the Universe are made. Scientists are trying to replicate this process on Earth to make a clean, safe and almost limitless energy source for mankind, but that's another story . . .

Fusion can only happen at very high temperatures because it is the forcing together of particles that would usually be very far apart. The nucleus is a minuscule dot in the space of an atom, about one hundred thousand times smaller than the atom itself (that's like a pinhead in the centre of a large football stadium). So even when atoms are bound together in molecules, the nuclei are, relatively speaking, nowhere near each other. Nuclei are also all positively charged particles, so when they are stripped of their electrons in the plasma state they still keep away from each other. Naturally, particles with the same electric charge, just like magnets of the same polarity, will repel. Nuclei don't want to come together, so the only way for fusion to occur is to heat things up a bit.

As you will recall, heating things up means that the particles will be moving faster. If they get up to speeds above about fifty thousand kilometres (31,000 miles) per second, they will be moving fast enough that the repulsive force won't act quickly enough to deflect them and the nuclei will collide. At the moment of collision, as they get close enough, the strong nuclear force takes over and pulls the new nucleus together. To ensure nuclear stability, anything unwanted is spat out. Protons, neutrons, electrons, neutrinos or light can all be emitted to balance the newly created nucleus. This is a fusion reaction. It creates vast amounts of energy as the nucleus rearranges, as if two colliding billiard balls had been fitted with an explosive charge that detonated on collision.

Stars are natural fusion reactors. The hot, dense conditions in the centre, created by gravity, are perfect for fusion. Stars start to form when little bits of gas and dust clump together, increasing the gravitational attraction of the clump (since the force of gravity depends on mass) and therefore pulling more matter into it. Gravity makes everything pull towards the centre, which becomes very hot and dense. Eventually the centre heats up enough that fusion reactions start and the star begins to shine. The energy released creates a pressure pushing outwards and balancing the gravitational force, pulling all the matter into the centre, so the star reaches an equilibrium that lasts until most of its hydrogen fuel has been fused away.

The temperature at the centre of the Sun is about 15,000°C and its density is about ten times that of lead. In the Sun, hydrogen fuses to helium in a slow, three-staged process of building up protons and changing particle identities (first spitting out an electron to change a proton into a neutron). This reaction releases light. Once all the hydrogen has been burnt, the star will contract again until the centre is hotter and denser than before and the bigger helium nuclei can be fused together.

Fusion reactions proceed in a sequence from small to large, just as if you were assembling building blocks or Lego bricks – small bricks are put together to make larger ones. In stars the small particles first fuse together to make bigger particles – hydrogen fusing to helium. Then the bigger particles fuse to make even bigger particles – helium fusing to carbon – and so on. In this way, bigger and bigger particles are built up in the centres of stars or, in the case of the heavy metals, in huge supernova explosions. From such humble beginnings, all the elements in the Universe were made.

The photon of light released in a fusion reaction in the centre of the Sun then begins its long journey out from the core, passing through the radiative zone and the convective zone before it reaches the solar surface. The photon can get trapped in the radiative zone for quite some time. It bounces off atoms at random like a puck in a pinball machine, being absorbed by one atom and then re-emitted to be absorbed by another. This seriously impedes its ability to exit this zone, which is the thickest layer in the Sun, stretching from around 20–70 per cent of the distance from the centre of the Sun to the surface – about half of the solar radius. It can take up to 200,000 years for a photon to cross the radiative zone travelling along this random, bouncy path.

Eventually the photon reaches the convective zone where the energy bubbles up to the surface like water boiling in a pan, travelling on currents in the gas. The outer edge of the convective zone is what we see, what we call the surface of the Sun – the photosphere. It's not a hard surface, nor is it smooth. Instead, it looks granular, almost bubbly, with small cells of convection bubbling matter up to the surface. These 'small' convection cells are about 1000 kilometres (600 miles) across. That's longer than the length of the British Isles – John o'Groats to Land's End – and that's just one tiny bubble on the Sun. By looking more closely at the Sun we see these features and realise that there is enormous dynamism at the Sun's surface – nothing is still.

Helioseismology – the study of the waves in the Sun – teaches us about its layered structure. Just as on Earth we can use the propagation of earthquake waves to tell us things about the medium through which they pass, on the Sun we can measure patterns of sound waves generated by turbulence at the top of the convection zone. The sound waves bounce around inside the star, penetrating to different depths and resonating like a musical instrument. Changes in sound speeds tell us about the density and temperature of the medium, so looking at acoustic wave patterns allows helioseismologists to build up a picture of the inside of the Sun – much like ultrasound for a star.

The temperature of the photosphere is about 5800K*. At this

---

* K is for Kelvin, named after Lord Kelvin who devised the absolute temperature scale, where zero is absolute zero, the lowest temperature that is theoretically possible. 0K is -273.15°C. The increments are the same as the Celsius scale. When we talk about very high temperatures, K and C are almost the same, e.g. 15,000K = 14,726.9°C.

temperature, light towards the yellowish-red end of the visible spectrum is emitted. The Sun actually emits light in all colours, but most of it is in yellow, so that is what we see. Light is also emitted outside of the visible range – light we can't see, like ultraviolet and X-rays. Looking at specific colours, or wavelengths, of light allows us to see different 'pictures' of the Sun. More specifically, it allows us to see its different temperatures. Hotter atoms give out higher energy light like X-rays or ultraviolet, so taking pictures in these wavelengths shows us the hotter atmosphere of the Sun. NASA's Solar Dynamics Observatory telescope has an instrument that takes pictures of the Sun in ten different wavelengths, allowing scientists to track how heat and particles move in and through its atmosphere. A collage of these images highlights the different views of the Sun that are obtained in different wavelengths.

It is here on the photosphere that we get sunspots – those dark regions of intense, twisted magnetic fields. Up close in visible light, sunspots look a bit like the Eye of Sauron in Tolkien's *The Lord of the Rings* trilogy, the bright granules of the photosphere puckering and stretching into long twigs that disappear into a central vertex shrouded in blackness. Sunspots come and go over days or weeks. These twisting magnetic fields also form loops that pop out of the Sun's surface, much like an elastic band twisted so tightly that it starts to kink and bulge. These loops, called prominences or filaments, drag red-glowing plasma with them out into the lower solar atmosphere and they can sometimes burst, throwing the plasma matter outwards. At times these overloaded magnetic fields cause explosive events that fling matter far out into the solar system; at others, during smaller bursts, the ejected matter simply falls back to the solar surface.

Let's go back to our fusion photon, which is now just leaving the photosphere. Although it has left the surface of the Sun, that's not the end of its story. The Sun has a hot and mysterious atmosphere, and its influence reaches out through the solar system via a solar wind.

The lower part of the atmosphere is called the chromosphere, named after the Greek word for colour because of its bright pinkish hue during a solar eclipse. It is a thin layer, only a few thousand kilometres thick on a star with a diameter of nearly a million and a half kilometres (that is like 110 Earths lined up shoulder to shoulder). In this narrow space the temperature rises by about 100,000K, so it is significantly hotter than the photosphere. It is also even less smooth than the bubbly photosphere, with larger cells surrounded by lots of spiky structures waving together like grasses in a breeze, revealing an underlying magnetic structure.

The origins of solar magnetism are complicated and still not fully understood, but we do know that it is to do with the movement of plasma (which, as you will recall, are charged particles, whose movement always generates a magnetic field). The total resulting field pattern arises from the interplay of convection in the convective region and the rotation of the plasma. It is similar to the generation of the Earth's magnetic field; however, it is complicated by rotation, which is not constant in the Sun. The central core rotates like a solid ball, with the poles and the equator rotating at the same rate. But the outer layers, from the convective zone outwards, display differential rotation, meaning that the equator rotates faster than the poles. There is a sharp region at the top of the radiative zone where the rotation changes. Scientists believe that within this thin layer is the origin of the solar magnetic field.

The solar magnetic field permeates from inside the Sun through the photosphere, the chromosphere and out into the Sun's outer atmosphere, the corona. The corona is hot, highly ionised matter around the Sun. In fact, it is intensely hot, mysteriously hot even – almost 2 million Kelvin rising from a minimum in the photosphere of around 5000K. It is invisible to the naked eye because it is so diffuse – and therefore faint – next to the bright disc of the Sun. But, like sunspots, mankind has nevertheless been viewing the corona since ancient times. It is visible during a solar eclipse, when the moon passes between the Earth and the Sun, blocking the Sun's light and revealing its wispy patterns.

The first inkling that the corona exhibited strange behaviour came from observations of an eclipse in the late 1860s. The spectroscope had just been invented, and scientists could now analyse light by splitting it up into its constituent colours, rather like a raindrop splits white light to make a rainbow. They soon noticed that different atoms emitted and absorbed specific colours of light and, because of this, by studying the light given off by an object they could infer its chemical composition. In the nineteenth century, using this new science of spectroscopy, several new elements were identified. Astronomers quickly set about experimenting with it. Already, Norman Lockyer had discovered a new element in the Sun. He named it helium after '*helios*', the Greek word for 'sun'. At that time this element had never been seen on Earth because, being lighter than air, it floats off into space as soon as it is released. It would be another twenty-five years before helium would be isolated on Earth. Lockyer discovered it in the Sun by examining the spectral line given off by solar prominences.

During the solar eclipse of 1869, Harkness and Young were independently examining the Sun's corona. They both saw a strong green line that didn't correspond to any of the colours that they recognised, and certainly not one they would expect from a temperature of 5000K, which was the acknowledged temperature of the photosphere. They thought it must be a new, unknown element. They named it coronium.

At that time, the understanding of heat transfer in the Sun was that there was a hot solar furnace at its centre, and that everything would cool as it moved outwards, continuing to cool out into space. But that wasn't the case. It wasn't until 1939 that the mysterious green colour – this spectral line – in the corona was identified as iron with thirteen of its fourteen electrons stripped off. This is a very rare atomic transition indeed, and one that can only happen at extremely high temperatures. The Sun was not cooling uniformly after all – the corona was hot! That high temperature is still something of a mystery, but it is generally accepted that the means of heating is to do with magnetic processes rather than more traditional means of energy transport. The corona is composed of hot, fast particles, so these particles are not confined to the Sun; they expand outwards and away – a hot, high-pressure plasma expanding into cold, low-pressure space. This expanding solar atmosphere is called the solar wind.

* * *

FOR THE LAST part of my Icelandic journey I met up with Felicity Aston, a polar explorer, and we went off in search of a radar called CUTLASS (Co-operative UK Twin Auroral Sounding System). In the tiny village of Thikkvibær we stopped by a field with a collection of what seemed to be large television aerials. This was in fact CUTLASS; it is one of a network of high-frequency radars known as SuperDARN that study flows in the ionosphere in the high-latitude regions – particle flows that occur when the solar wind hits us and is deflected around Earth. The radars measure the level of disruption in the upper atmosphere, which can be correlated with magnetic measurements and data from spacecraft missions better to understand how the Sun and Earth systems move and interact.

As we drove, Felicity told me some of her experiences of the northern and southern lights. In her twenties, she spent three years living and working in Antarctica as a meteorologist with the British Antarctic Survey. Though she was there during the winters, she was based at Rothera Station, on the opposite side of Antarctica to the South Magnetic Pole. Felicity only ever saw the aurora once while she was there, a green smudge on the horizon on her twenty-sixth birthday. Since then she has been on numerous polar expeditions and, in 2012, became the first woman in the world to ski across Antarctica alone – 59 days from the Ross Ice Shelf to Hercules

Aurora australis (the southern lights), seen from the International Space Station.

The church in Kiruna, Sweden.

On the sled behind Knut's snowmobile, off to find the reindeer herd. Karasjok, Norway.

A view across the fjord from Alta to Haldde mountain, Norway, where Kristian Birkeland overwintered in 1899 to make auroral observations. A larger, more permanent observatory was erected later and it can be seen in the picture as the small, rectangular bump on the central slope.

Rhyolite colours in the Landmannalaugar region, Iceland.

Standing next to a steam vent along the Laugavegur trail in the volcanically active region of Landmannalaugar, Iceland.

Aurora over the water off the Vatnsnes Peninsula, northwest Iceland. In the foreground is the 15m basalt stack of Hvítserkur.

A drawing of sunspots on the surface of the Sun, made by Galileo in 1612. Though not the first person to observe and sketch sunspots, he is certainly the most famous and is also credited with having coined the phrase ‘aurora borealis’ (meaning ‘northern dawn’) in reference to the northern lights.

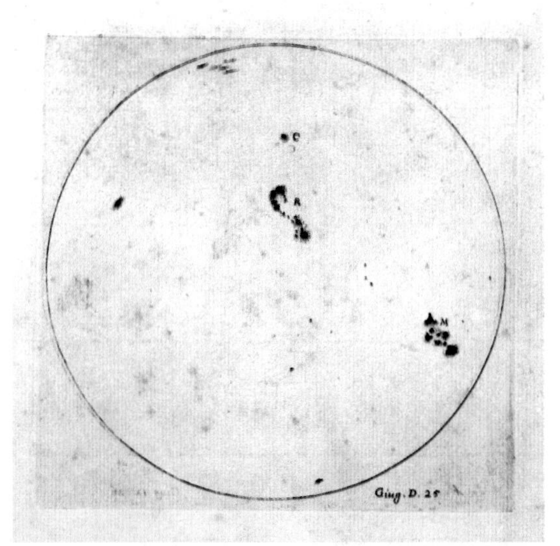

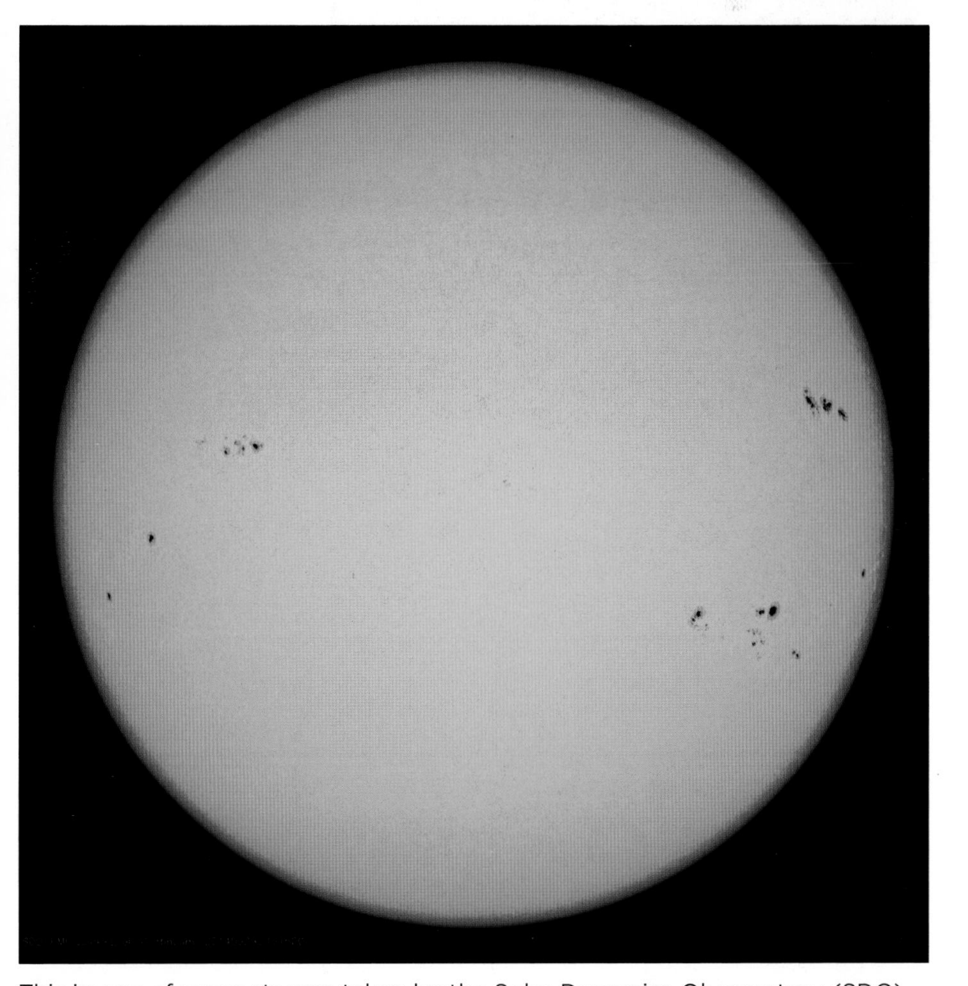

This image of sunspots was taken by the Solar Dynamics Observatory (SDO) satellite by the HMI Continuum camera on 29th September 2014.

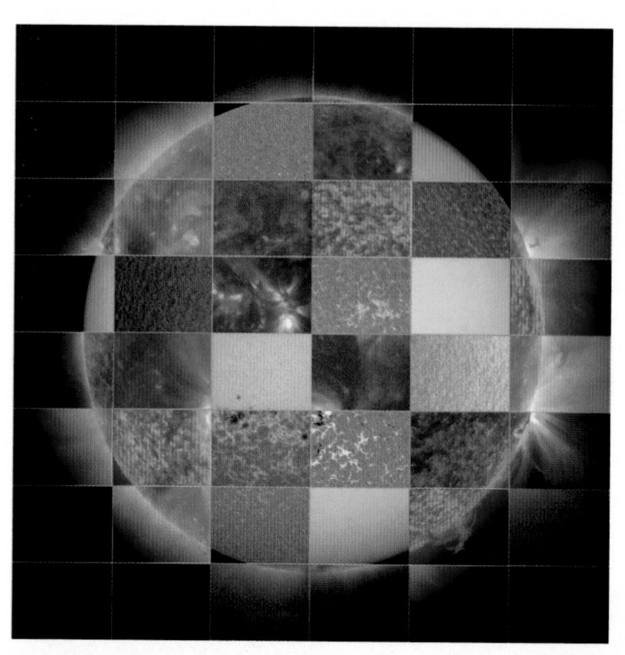

This collage of solar images in different wavelengths is from NASA's Solar Dynamics Observatory (SDO). It shows how observations of the Sun in different wavelengths allow us to see different features of the Sun's surface and atmosphere. The collage also includes images from other SDO instruments that display magnetic and Doppler information.

The CUTLASS radar in Thikkvibær.

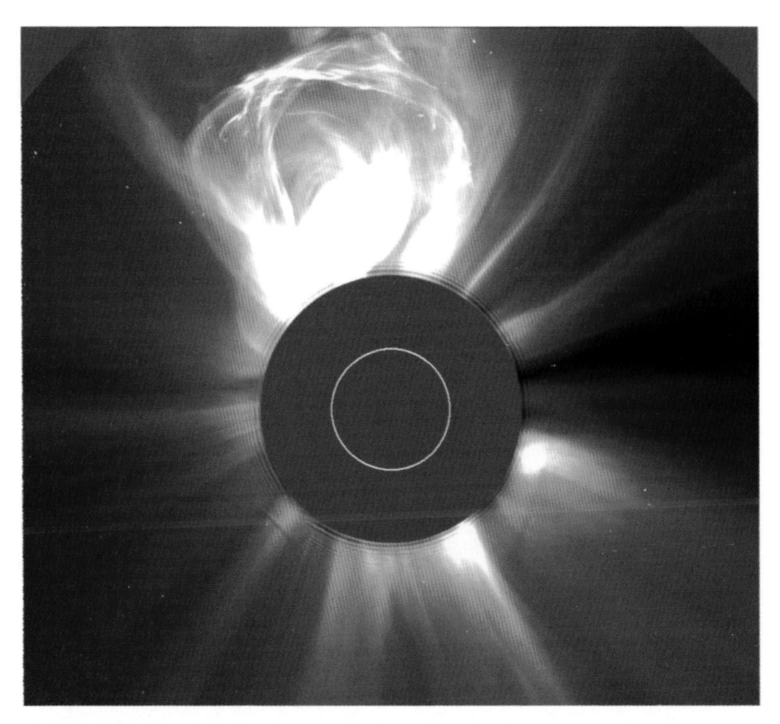

Image of a coronal mass ejection taken by the coronograph instrument on the SOHO satellite. The size of the Sun is marked by the inner white circle. Billions of tonnes of matter are thrown out into the solar system at several million miles per hour.

An early view of the entire auroral oval in ultraviolet, taken by the NASA Dynamics Explorer 1 satellite on 8th November 1981. A map is superimposed on the image and the glowing oval of auroral emission is about 4500 km (2800 miles) in diameter.

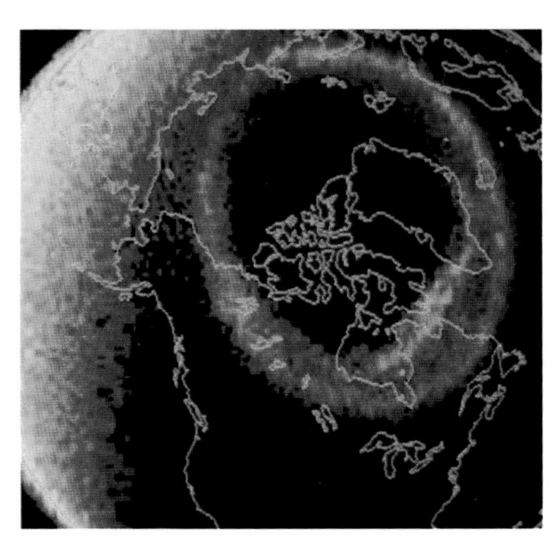

What the camera sees versus the human eye. A photograph (top) of the aurora, taken in Finland, and the same photograph (bottom) adjusted to show how the photographer remembers seeing the scene. © Chris Lee

Inlet via the South Pole. But these expeditions were during 24-hour daylight; it is during the winter that the aurora brightens up the dark polar skies.

My journey is focused on the northern lights rather than the southern lights, not only because the Arctic is so much more accessible than the Antarctic, but because there is human habitation and therefore history, culture and stories surrounding the northern lights. Yet I hope one day I'll get to Antarctica, too.

Felicity now lives in Iceland, so she often sees the northern lights when the weather is clear enough. It is possible to see them right from the centre of Reykjavík, though driving out of town makes for much better viewing. She is impressed at how the locals don't seem to tire of them. 'People here have seen the aurora every winter of their lives, so it could be quite an everyday occurrence,' she said. Felicity told me that often when she drives out of Reykjavík to get a better view of the display she is amazed by how many other cars are there – fellow Icelanders going out specifically to appreciate the lights. 'It's something quite special,' she said, 'maybe because it's never quite the same.'

The volcanic nature of Iceland can make for an even more spectacular experience. Felicity described to me a time when she saw the aurora alongside an eruption in the Icelandic highlands. 'The volcano was turning the sky these amazing pink and orange colours, and there was a layer of cloud caused by the eruption that was being lit in lots of different colours. Above that was a clear sky with so many stars out. Then we started to see the green bloom of the aurora.' Felicity, like others before her, felt as if this was something belonging to another planet. 'It was really surreal to see so many colours from two different sources in the sky,' she said. 'It's just quite otherworldly, like something from a sci-fi film showing what the sky of Saturn might look like.'

As well as being a geological showpiece, Iceland is in a prime spot for viewing aurora. The pattern of the Earth's magnetic field often brings the aurora to right above the Icelanders' heads, stretching out east over northern Scandinavia and Siberia and west over Canada and Alaska. The next part of my journey took me west, to Canada, to discover what happens in our magnetosphere and our atmosphere to generate the bright, mesmerising auroral displays we see in these high-latitude places.

When I returned to Reykjavík after my gritty hike in the Laugavegur and my day with Felicity, I met a physicist, Gulli Björnsson, for tea. He told me that his predecessor had been running the Leirvogur Observatory at the University of Iceland for fifty years before he retired, and that this man was adamant that the most impressive auroral displays occurred two to three years after solar maximum. It seems that at this time in the solar cycle there are often some large coronal holes near the equator releasing fast solar wind which buffets the Earth for a number of days and creates prolonged and vibrant lights.

As we spoke, I made a mental note to return to Iceland in a few years' time. In winter.

CHAPTER FOUR

CANADA – ELECTRICITY AND MAGNETISM

I STOOD ON THE ladder, my face centimetres from an SLR camera mounted sideways, its lens pointing straight up to the sky. I followed its gaze and tilted my head back to look up through the dome, straight up into the blue blue sky of northern Canada in summer.

I was in Athabasca, a small town in Alberta, Canada, about four hours' drive north of Calgary. The town lies at the southernmost bend of the Athabasca River, where it turns back towards the north after a meandering detour, and for a few short decades after its founding in 1877 it was a boom-town staging post – the 'Gateway to the North'. Back then it was known as Athabasca Landing, the terminus of the Athabasca Landing Trail that linked the North Saskatchewan River system with the Mackenzie River system, of which the Athabasca River is a part. The trail was the major route to the Arctic for fur traders, gold-rush miners, missionaries and settlers. For a brief period in history, Athabasca boomed; however, by 1919 the development of the railways had reduced the town's importance as a trading post. Instead, lumbering and farming, alongside oil and gas explorations, became the staple occupations. Now the quiet town hosts a university known for its distance-learning courses – it is the Open University of Canada – and for its small observatory. Well, two, in fact.

The original Athabasca University Geophysical Observatory opened in 2002 on the main university campus. Within a decade a new, larger facility was built 25 kilometres (15 miles) outside the town as the research activities expanded and the scientists sought darker skies. The old observatory now hosts the magnetometers that measure variations in the geomagnetic field. There is also a telescope, hidden beneath its protective white, clamshell dome, but most of the imaging is now done from the new, darker observatory.

I spent a night there under the supervision of Martin Connors, the observatory director. Serious and softly spoken, he told me that specific developments in town had caused the light pollution problem, particularly a new sports centre with a large and well-lit car park. It is not too big an issue for the telescope, which focuses on specific objects, but the washed-out light in the sky makes viewing of diffuse phenomena such as the aurora difficult. To make matters worse, at Athabasca their research speciality is proton aurora, which is the faintest kind and blue in colour, making it very hard to discern. (The vivid aurora we see and talk about most often is caused by electrons.) So the observatory moved out of town, and it gave them an opportunity to increase the facilities.

After a simple dinner in Athabasca and a walk to the famous river, we drove out to the observatory at sunset. Fairly shortly after leaving town we turned off the main tarmac road onto dirt tracks through forest. We passed a long lake, still and glassy and hung with a fine layer of mist, which edged the lake in blended greys as the centre softly echoed the hues of the sky. The horizon was still pink as we drove up to the observatory compound, the building encircled by a high chicken-wire fence.

The new Athabasca University observatory looks like a big white rectangular box with small bubbles on top. These are the clear acrylic domes used for optical observations. The building is functional and unadorned, except for some colour provided by a low band of taupe and wide stripes of dusty Air-Force blue up the sides of the box like ribbon tying a present. The ground floor sports a row of white-framed windows but the upper floor is conspicuously blank. The only light getting into those darkened first-floor rooms is that coming in via the observing domes. Inside there is a large, open hallway that doubles as a workshop, four small bedrooms where researchers may sleep, and a kitchen area for meals and relaxation. It is basic but comfortable. The next day, Martin would take me up on to the roof to see the domes and show me the instruments and cameras inside them, set up for a fisheye, all-sky view from horizon to horizon.

That evening, though, we walked out from the observatory through the wild grasses and low forest. I followed Martin down the slope and along the stony, muddy path heading through the trees. The bugs wheeled around us, biting me in my hair where I didn't have repellent.

'I can't believe how light is it,' I said. 'It's twenty past ten!'

'It will nevertheless get darker,' Martin replied, 'but it will still be quite light till midnight.'

Daylight changes rapidly at higher latitudes. Already past the summer solstice, the days were steadily shortening. 'By September we will have twelve-hour days and twelve-hour nights, then by December we will have lots and lots of night,' said Martin, 'which is an advantage for the aurora.' Winter days are little more than seven hours long in Athabasca, which gives long nights for observing. The best viewing conditions are in deep winter – January–February time – when the polar air masses move in, the temperature drops to below -20°C and the sky is very clear.

'Does it often get that cold here?' I asked.

'This past winter it got down to minus 40 for the first time in several years. We had not seen minus 40 for at least the previous three, but it was fairly routine before that.'

Low temperatures might be good for viewing, but they are not so good for the telescope and the opening mechanism of the dome, which tends to freeze up if the temperature drops below -25°C. Fortunately, getting too cold is not a problem for the cameras in the new, heated observatory. Martin has more trouble keeping them cool in the summer, with the long hours of intense sunlight shining through the transparent domes. It is like the instruments live in a greenhouse and the air-conditioning system has to work hard to keep them cool.

In the silence our footsteps crunched loudly and our voices hummed through the trees, scaring away any wildlife – though we did keep a lookout nonetheless. Martin spotted some paw prints in the mud of the path.

'That, I think, is just a dog,' he said, examining the strongest impression in the ground. 'See, it's got claws?' He pointed them out to me. 'Cougars don't have claws that show.'

'Okay,' I said, curiously. I knew nothing about cougars, except that they were some kind of big cat. How big, I had no idea.

'Oh, I forgot to tell you,' Martin continued, 'there are cougars here.'

I laughed nervously. 'So cougars as well as bears?' I asked rhetorically.

'Cougars will usually only go for a single person. They won't attack two people.'

'Do they actually attack people?' I asked in disbelief.

'Oh sure, but they want to know they're going to win.'

I later found out that cougars are also known by the names of mountain lion, puma and panther.

We walked on past a small shooting range and emerged out of the forest to the lake shore. There was a small speedboat moored up on a wooden jetty and a couple of kayaks pulled up on the grass. This was an Air Cadets' camp, Martin told me. As we approached, a dark animal slid softly into the water – a small beaver or a muskrat, perhaps. I looked out across the lake. The mist was still rising and in the twilight now the colours had darkened to washed-out indigos and pale violets.

Besides the kayaks lying out in the grass there was no sign of life at the camp. The place was deserted. Martin pointed out the mess hall, the director's house and the cabins where the children stay. In the winter Martin comes to give astronomy talks to the Air Cadets and takes them out stargazing, using a laser beam to point out the constellations he had been telling them about earlier.

We walked back to the observatory, chatting about wildlife and weather and stargazing. As we emerged into the clearing and walked back up towards the building, Martin pointed out some of their unusual garden ornaments – strange, spindly sculptures made from wire.

'That's a VLF antenna,' he said, pointing to a tall pole with two roundish loops at the top and blue guy lines stretching to the ground. 'It doesn't transmit, it just receives.'

The antenna receives background Very Low Frequency radio-waves which are given off by particles in the magnetosphere, so the signals can tell us what is happening in the plasma out there.

'And that thing that looks like a chicken pen is a riometer. It's detecting galactic noise, but if there is a lot of absorption going on in the ionosphere then the signal from the galactic noise cuts out.' It is the aurora that disturbs the ionosphere so that it absorbs the galactic noise, so the riometer signal dropping indicates a sudden brightening of the aurora. As well as the all-sky cameras, optical instruments and the magnetometers, these antennae form the toolset necessary for studying the aurora.

The unassuming building where I stayed in Athabasca is one of the most accessible observatories for auroral research. As Martin said, they are at the edge of the auroral zone but also at the edge of civilisation. Alongside its dark skies it has reliable power and water infrastructure, high-capacity internet and good transportation. It's a smooth, two-hour drive from Edmonton, with its international airport and surprisingly gigantic shopping centre. It is North America's largest mall, which happens to contain – besides shops – an ice-rink, swimming pool with waves, a cinema, an amusement park with rollercoaster, a sea-lion show and a 'European street', as well as possibly more delights that I didn't see. Having heard about the temperatures the people experience up here, it wasn't surprising that they wanted to keep everything contained.

Athabasca may be exceptionally accessible but, in fact, the whole of Canada has an advantage over the rest of the world in terms of ground-based auroral studies. The geomagnetic axis is tilted by about 10 degrees towards Canada from the Earth's rotation axis. In other words, the North Magnetic Pole is offset from the geographic pole.

We know from our earlier travels that it is the geomagnetic field that is fundamental in the creation of the aurora, so auroral activity occurs in a ring (or, in fact, an oval) centred on the magnetic poles. The tilt of the magnetic axis towards Canada in the northern hemisphere means that the oval reaches to lower latitudes in North America than on the other side of the planet. Athabasca is at a geographic latitude of almost 55°N, which is about the same as the city of Carlisle in the north-west of England, near the border with Scotland. However, its geomagnetic latitude is 61°N, whilst that of Carlisle is only 57°N. The United Kingdom is only about a quarter of the way around the planet from Canada so it doesn't suffer the full degradation in latitude of the geomagnetic tilt that Asia does, situated directly opposite Canada. The city of Petropavl in Kazakhstan, close to the Russian border, is almost directly on the other side of the world to Athabasca and at nearly the same geographic latitude. Petropavl has a magnetic latitude of only 47°N. The chances of her citizens ever seeing the northern lights there are almost non-existent. Residents of the southern parts of the UK seldom see auroral activity, though it becomes more likely further north in Scotland, whilst in Athabasca it is frequent enough to site an auroral observatory.

The vast majority of land situated in the auroral zone belongs to Canada. The auroral zone is the land above which we generally see the aurora, by definition at midnight. It is a band demarcated by magnetic latitude, stretching approximately 1300 kilometres (800 miles) between around 61°N and 73°N magnetic, based on probability, though the *International Auroral Atlas* defines the zone as between 60°N and 75°N magnetic. The auroral zone is subtly different to the auroral oval, though sometimes when describing night-time aurorae they are used interchangeably. The auroral oval is the instantaneous spread of the aurora around the globe, which opposite the Sun roughly coincides with the auroral zone, but which on the day-side is considerably narrower and further north.

The confusion probably arises because the meaning of the name 'auroral oval' has changed over time. It was first coined when people started making detailed observations of the aurora from the ground in the nineteenth century. From the ground we cannot see the whole of the auroral oval because aurorae are predominantly at around 100 kilometres (60 miles) altitude, so the horizon obscures anything happening more than about 1000 kilometres (600 miles) away. Early studies of the aurora involved amassing statistics from reports of sightings of the northern lights. Scientists worked out where and under what conditions they most often saw the aurora. They obtained probability measures associated with different conditions of measured magnetic perturbations on the ground. This gives a ring of greatest probability of seeing the aurora. The first auroral ovals were lines on maps indicating where one was most likely to see aurorae, in other words, the auroral zone. It wasn't until the space age, when satellites went up and started taking pictures of the Earth, that scientists realised that the auroral oval was not simply a probabilistic construct – we actually do see a continuous ring of aurorae around the Earth.

Now think about the distribution of the auroral zone, this twelve-degree cincture where auroral likelihood is highest. Visualise it as a thick band of colour drawn in a ring across the globe. It sweeps down through Alaska and dips across the wide landmass of Canada. Once it crosses Hudson Bay and northern Quebec and leaves North America, it traverses the sea to the

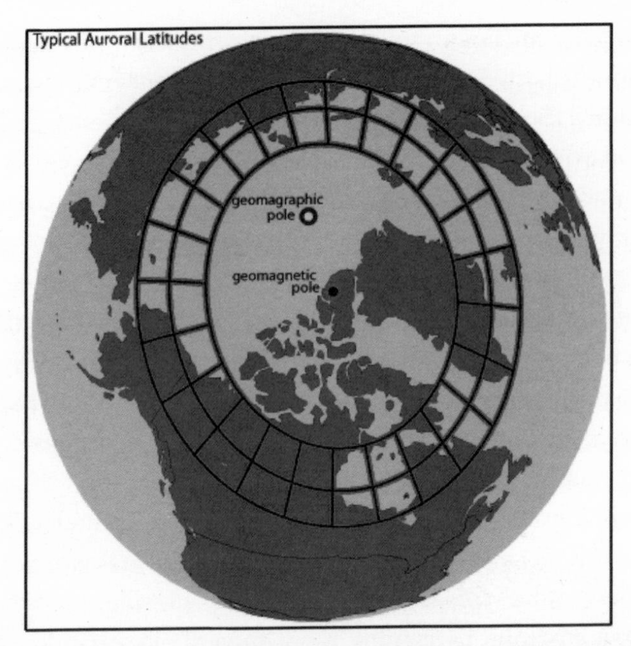

Map of the Arctic showing the auroral zone. Lines marked are 65, 70 and 75 degrees magnetic latitude. Canada has the bulk of the land in the auroral zone. Courtesy of Eric Donovan of the University of Calgary.

southern tip of Greenland, then to Iceland, across the Norwegian Sea into northern Scandinavia, then across the vast coast of Siberia back to Alaska. By far the majority of land covered is Canadian, and this has shaped their science programmes and the way they study the aurora. With all this populated landmass in the auroral zone, Canada is a perfect place for ground-based imaging. Scientists have taken advantage of this, creating a network of small cameras across the country.

As with other northern nations, the aurora has been linked with Canada for her whole history. The name 'northern lights' has been used for things as wide-ranging as a night-time classical music radio show, a municipal district in Northwest Alberta, a 1980s Canadian celebrity musical group (the Band Aid equivalent that raised money for the famine in Ethiopia) and a blended Canadian whisky. They even had a wartime comic book superhero (superheroine, in fact) in the 1940s called Nelvana of the Northern Lights. In her fur-trimmed miniskirt, knee-high boots and matching gloves and cape she would fly at the speed of light on a giant ray of the aurora, protecting the Arctic North. So it seems natural that Canada would also have strong scientific research activities centred on the northern lights. And indeed it is so. It is a geographical advantage which allows them to increase their relevancy in the scientific world.

Before travelling north to Athabasca, I spent several days at the University of Calgary in the company of Eric Donovan, Professor in the Department of Physics and Astronomy. We talked in beautiful meanders through scientific history, auroral observation, plasma physics, satellite missions and outreach before I set off up Highway 2. I spent my days in Calgary at the Physics Department, a short walk from my hotel across the grassy campus. I'd go back to the hotel each evening with my head bursting with information, sometimes stopping on the way back to lie down on the grass in the sun and reflect on the day's discussions. Eric was jovial, with a ready smile, grey-brown hair and beard and a casual demeanour. He was passionate and interesting to talk to, always wanting to tell me more, and enthusiastic about the Auroral Imaging Group's work with smaller cameras.

'When thinking about the history of science,' Eric said, 'one should always think about why people do what they do.' It is tied in with the social history of the country, indeed the world. The obvious example is the surge in understanding of nuclear physics during the Second World War, but there are many others besides. Eric cites the investigations into thermodynamics during the industrial revolution. Even Kristian Birkeland, as a side project to fund his personal passion for auroral research and his expeditions, invented a process for making fertiliser because it was clear that there was an economic need. On examining the history, we can better understand where the funding comes from for the research.

Towards the beginning of the nineteenth century, nations began to recognise that a good scientific output brought recognition on the world stage, which in turn attracted investment and activity within the country. Nations began to desire to be scientifically relevant in the world. For the big, powerful countries this was easy – they had the resources to work on whatever they wished. The smaller countries with more limited resources, however, had to think more strategically if they were going to gain recognition alongside their larger neighbours. They had to make sure that what they did made sense within their environment. Consequently, the aurora became a great research interest in Scandinavia and Canada. It was clear that the aurora was a natural phenomenon, and the interesting problem had been perplexing scientists for centuries. Moreover, it was clear that the emerging technologies of photography and spectroscopy could be used to study them, but one needed to be in the auroral zone to do so.

Yet even between Scandinavia and Canada, research approaches were different. Canada has the vast majority of land in the auroral zone; Scandinavia has a mere toehold in comparison. So in examining their narrow slice of the auroral oval the Scandinavians use many high-level instruments to make detailed measurements on this smaller scale. They often have state-of-the-art cameras to study dim or fast-changing aurora; or they use tight networks of cameras to perform triangulation or tomography to determine height distributions. They operate the EISCAT (European Incoherent Scatter Scientific Association) radars – powerful systems used to study disturbances in the ionosphere noticeable by their imposing, parabolic dishes 32 metres (104 feet) across. Conversely, Canada, with all its auroral zone land, has the ability to cover a range of spatial scales. They look at the mid-scales with fairly cheap instruments spread over a large region. Funding has been put into setting up the wide infrastructure, stretching from the east coast at Goose Bay to Inuvik in the far north-west, and managing the large amounts of data that are generated. So the approaches are different – complementary – but they are shaped by the landscape and geographical advantage. As Eric says, 'find your niche, be good at it and then you can be relevant.' Canadian scientists are playing to their strength.

Due to the chaotic movements of fluid in the Earth's outer core, the geomagnetic field is changing and the magnetic poles are drifting slowly over time. In 2001 the North Magnetic Pole was measured to be situated at 81°N 111°W in the Arctic Ocean, to the west of Ellesmere Island. When Amundsen skied to it in 1903 he measured it to be at 70°N 96°W, much further south along the route of the Northwest Passage. Its location was not much different when it was first measured by the Briton James Clark Ross in 1831, but over the past century its wandering has been accelerating. The magnetic pole is now moving northwestwards at about 55 kilometres (34 miles) per year. Regardless, with the large landmass of Canada stretching so far north, albeit into more hostile Arctic regions, the dominance of Canada in the auroral zone is relatively assured.

The camera mere centimetres from my face in the dome of the observatory at Athabasca was a commercial CCD-based digital camera, modified with state-of-the-art lenses by the team at the University of Calgary. There are tens of these observing stations with similar cameras dotted around the large country. The camera was of a type often used by amateur astronomers and the additional optics were topped by a fisheye lens to give a wide, circular field of view. Pointing directly upwards, the camera looks out in a cone from horizon to horizon, which is why it is called an all-sky imager. Looking at an all-sky picture, around the edge one can see trees and other objects on the horizon poking into the circle. Up at an altitude of around 100 kilometres (60 miles) where the aurora is seen, the field of view is about 1000 kilometres (600 miles) across, so when the images of several cameras are merged together they form a mosaic of aurorae across the entire continent. The cameras turn on at nautical twilight and transmit data via satellite in real time.

Looking directly up at the sky through the rooftop domes, the all-sky cameras collect images every six seconds that show the beautiful arcs of the aurora – streamers of light across the sky, stretching, brightening and breaking up. Continually imaging throughout the night, spectacular time lapses are built up so that we can watch the evolution of these auroral storms. But we can do more than just watch. The light that we see in each little pixel of the sky corresponds to a point on the magnetic field line of the Earth that stretches up into space. Looking at the shapes and patterns of the aurora allows us to look up far into the atmosphere and beyond along the field lines, up to where the bubble of the Earth's influence meets the plasma stream coming from the Sun – the expanding solar atmosphere, the solar wind.

* * *

THE SOLAR WIND stretches out from the Sun into space, pervading the entire solar system and taking the Sun's magnetic field with it. We call this the Interplanetary Magnetic Field. It eventually hits plasma from interstellar space, impacting in a shock front because magnetic fields can't mix. The bubble of space dominated by the solar wind is called the heliosphere, and the Earth and the other planets sit out in the flow. The heliosphere received some attention recently because in 2012 the Voyager 1 spacecraft finally passed the boundary of the heliosphere and went into interstellar space, after about 35 years of flight.

The magnetic field of the Sun transfers out to the atmosphere and the solar wind because of an interesting property of plasmas often described as the 'frozen in' condition. This demonstrates that the magnetic field and the bulk plasma move together. Usually one dominates, so either the magnetic field drags the plasma around or vice versa. Think of a string of pearls or beads lying unclasped on a table. Each pearl is a blob of plasma and the string is the field line. If you were to gently pull a pearl or two sideways then the string within would come as well, like a field line moving with the plasma. Conversely, were you to pull only on the string, the pearls would move with the string, like plasma moving with the field. It is for this reason that the magnetic field of the Sun travels out through the solar system on the solar wind. The magnetic field is generated in the Sun and frozen into the plasma, remaining with it even as hot plasma escapes from the surface. This 'freezing' only breaks down under the most extreme conditions, such as when the magnetic field varies greatly over a small distance in plasma turbulence. This can cause sudden magnetic breakdown and re-arrangement, then the plasma is frozen with the new magnetic field. This breakdown and subsequent 'reconnection' to a new magnetic configuration is a vital part of the story of the aurora.

If you look at the Sun during a solar eclipse you will see the corona and the magnetic field patterns within. What you see is unique because the solar magnetic field is always changing, so the detailed pattern will never be the same again. However, broadly speaking, the solar field is dipolar like the Earth's, so we see the magnetic field lines diverging up and outwards like rays at the poles and curving more strongly into the dipole, butterfly-wing pattern as we move slowly equatorwards. What is different about the field pattern of the corona is that it is being stretched *away* from the Sun. As the solar wind flows away, it takes its frozen-in field with it, stretching out the dipolar magnetic field to look like a pinched loop. Since the magnetic field has a direction, with a north and south polarity, far away from the Sun where the field lines are pinched together there will be a hard division between north and south.

Imagine cars going around a simple race circuit. They are moving in a circle, a loop, all going in a single direction. Now imagine stretching out the circuit so it is an ovular shape. The cars still look like they are following a loop in a particular direction. Now imagine stretching it out so far – perhaps for many kilometres – so that at one end there is a tight hairpin bend. Apart from at the bend, anywhere along the stretched parallels the cars will be travelling in opposite directions, like on an ordinary road.

On the Sun, this stretched road is the magnetic field travelling out with the solar wind. The oppositely directed cars are the north and south polarities, and the central reservation is a sheet of moving charged particles, a current that encircles the Sun like a stiff ballerina tutu stretching out into the solar system. This ballerina skirt moves – it rises and falls – meaning that the Earth sometimes sees beneath the skirt to oncoming plasma of one polarity, and sometimes above, to plasma of the opposite polarity. Thus the direction of the magnetic field of the solar wind hitting Earth can change from southward to northward.

Of course, the Sun's magnetic field does not stay constant and the Sun's rotation, magnetic loops, coronal holes and sunspots all interrupt the dipolar picture, which also causes the magnetic field direction in the solar wind to change. This can have effects for us on Earth. The Sun's activity waxes and wanes on an eleven-year cycle, transitioning from a quiet, low-activity period (solar minimum) to a heady, twisted, sun-spotty, high-activity period (solar maximum) and back. Around solar minimum, the magnetic field is broadly the stretched-out dipolar field as described above, but the field gets particularly messy around solar maximum. Over the course of the solar cycle, more sunspots appear, and they move gradually towards the equator. Being intensely twisted loops, sunspots have a strong magnetic field and can pull the magnetic axis (the butterfly body in the dipole butterfly-wing picture) down towards the equator. This mixes up the field pattern, and the north and south polarities, and eventually the Sun's total magnetic polarity will change – north becoming south and south becoming north. This flip marks solar maximum and it happens roughly every eleven years or so. These messy fields at solar max are the cause of more frequent disturbances, and auroral activity, on Earth.

Around solar maximum there is also more chance of seeing other explosive events that happen in the solar atmosphere – coronal mass ejections, also known as CMEs. These occur when the closed-loop magnetic field above a sunspot becomes so intense that the field eventually breaks and reconfigures, throwing out matter into the solar system in a giant blob. Coronal mass ejections are big: they can contain billions of tonnes of matter accelerated to several million kilometres per hour. When viewed using a coronograph (which is a camera that makes a fake eclipse by blocking the bright Sun with a metal disc so that the faint corona can be seen), we see a huge tongue of plasma licking out into the heliosphere.

All the trapped magnetic energy associated with sunspots can also be released as a solar flare – an intensely bright burst of light in almost every wavelength of the spectrum. Solar flares are classified by the amount of X-rays they give out, and they often – but not always – trigger coronal mass ejections. The first solar flare identified was by the British astronomer Richard Carrington in 1859.

Carrington was a gentleman astronomer, second son of a wealthy brewer, who built his own private observatory in Surrey, England. On 1st September 1859 he was drawing sunspots from an image of the Sun projected onto a small screen. He saw the Sun as a simple white circle, slightly darkening at the edges, and the shadowy sunspots on the surface. Although viewed in greyscale, he made detailed drawings of the irregularly shaped sunspots with detailed shading of the light and dark patches. As he worked that day, suddenly he saw 'two patches of intensely bright and white light' break out. He was surprised, thinking at first that there was a hole in the screen because 'the brilliancy was fully equal to that of direct sunlight.' He noted down the time.

Another amateur astronomer, Mr Hodgson, who was also observing sunspots on the other side of London, corroborated this unusual sighting. It was the first observed solar flare. It was also the first time that a cautious connection was made between activity on the Sun and effects on Earth. At the same time that Carrington had witnessed the solar flare, magnetometers at the Kew Observatory in London had recorded a strong disturbance in the Earth's magnetic field, and the following day the world experienced one of the biggest magnetic storms ever recorded. Dazzling displays of the aurora were seen as far south as the Caribbean. The Sun had released a huge coronal mass ejection at the same time as the flare, which had travelled directly to Earth. Carrington suspected a connection between the flare and the aurora, but was nevertheless quick to caution that more examples would be needed, remarking that, 'One swallow does not make a summer.'

Solar flares and coronal mass ejections are extreme cases of how stored magnetic energy can be released, but even low-level activity has effects on Earth. On the Sun, nothing is constant. It is rotating, gases are moving, the magnetic field is changing, and we, the humble Earth, sit in this turbulent particle pool. It is just the geomagnetic field and the atmosphere that protect us. The solar wind flows around us as though we were a pebble in a stream, but it doesn't simply wash over us: it affects us too. The direction of the solar wind field determines if and where solar particles can enter the Earth's domain, which we will discuss later; and the energy of the solar wind determines the scale of the effect.

The solar wind can have different energies, or speeds. Fast solar wind comes from coronal holes – regions on the Sun where field lines open straight out into the solar system, so plasma flow is unrestricted. It is often seen coming from the Sun's poles, especially around solar minimum when the magnetic field is relatively dipolar. As the fields get mixed up around solar maximum, so do the fast and slow solar wind streams. A couple of years after solar max, there are often a few large coronal holes found towards the equator. Fast solar wind particles travel at about 800 kilometres (500 miles) per second, whilst slow solar wind travels at around 300 kilometres (190 miles) per second, but it's not just the speed that varies; fast and slow solar wind also has different compositions because they originate at different depths in the solar atmosphere. The slow wind is also denser than the fast, so there are more particles in a particular volume packet, but that is still not much. There are about ten particles (protons and electrons) per cubic centimetre in slow wind. Even though this is more dense than fast wind, it is still fewer particles per centimetre cubed than we can ever achieve in even the best vacuum on Earth. Yet although the solar wind is extremely rarefied by Earth standards, it is still able to cause significant disturbance on Earth, and it is the faster solar wind particles that create some of the most beautiful aurora here – those that Gulli Björnsson's predecessor in Iceland maintained occurred two to three years after solar maximum.

* * *

OVER THE PAST century, photography and imaging have revolutionised study of the aurora, allowing form and structure over time to be captured and stored, to be reanalysed at a later date.

Though the roots of photography stretch into the Renaissance period, progress in actually producing a permanent image did not occur until the mid-nineteenth century. These roots could be described as the knowledge that sunlight causes salts of silver such as silver chloride to darken, or that a black box with a pinhole opening (known as a camera obscura) could be used to project an image onto a back screen. But until the 1800s, the images in the camera obscura were too faint to darken the silver, and even if sunlight and stencils were used, there was no way of 'fixing' the images. They would degrade in daylight.

It was the French inventor Joseph Nicéphore Niépce* who produced the first permanent photograph, in 1827. It was a picture of the courtyard outside his workroom and it was captured using a camera obscura, imprinted over eight hours or more in a layer of sun-hardened bitumen on a pewter plate. Niépce's interest in photography began with lithography, at that time a novel printing technique, but not being artistic himself he struggled with obtaining the base image for printing (his son helped with this until he left for military service). Thus Niépce was inspired to seek a way to reproduce images from nature.

---

* Niépce is pronounced Neep-sea.

Another Frenchman looking to reproduce images from nature was Louis-Jacques-Mandé Daguerre, an artist and set designer at the Paris Opera who co-founded the Diorama theatrical experience. This used semi-transparent linen sets painted on both sides and manipulated light to achieve changing seasonal or diurnal effects. The Diorama was a sensation in the 1820s in Paris and London, but Daguerre wanted to take it further. He wanted to include real-life representation and had been attempting – unsuccessfully – to 'fix' images on silver chloride paper. He heard of Niépce's success through mutual acquaintance (the Chevalier brothers, the opticians who had made Niépce's camera obscura) and they met in 1827. Initially in partnership and later, after Niépce's death, alone, Daguerre built on Niépce's work. He used iodine to increase the sensitivity of the silver coating and mercury vapour to develop, or bring out, a very faint image from a short exposure. In 1839 his daguerrotype method was declared and others began producing individual photographs on silver-plated copper.

At the same time that Daguerre was working on his method in France, William Henry Fox Talbot was also establishing his own method of photography in England. He used the light-sensitive silver chloride in paper, discovering that the light sensitivity could be reduced with common table salt, thereby fixing the image and making it permanent. In this way he produced negatives, his earliest in 1835. The fixing process was improved by Sir John Herschel, who in earlier years had discovered that salts of silver would dissolve in a solution of sodium thiosulphate (hyposulphite of soda). This proved better than salt. In 1840, prompted by Daguerre's method of using mercury vapour to bring out a latent image, Talbot discovered that gallic acid could increase the strength of an invisible image imprinted into the silver salt. This effectively increased the speed of his photography, so his exposures could take minutes rather than hours. So the foundations of photography were laid.

Over the remainder of the nineteenth century improvements were made to the photographic process and to the plates, and later films, increasing the resolution and reducing exposure times. Lenses were also improved, and colour photography was achieved by James Clerk Maxwell in 1861, though a successful commercial solution would have to wait until the turn of the twentieth century. The biggest improvement was the use of so-called emulsions (they were not true emulsions in the chemical sense of the word, that is, a suspension of a liquid in another liquid, but the term endured nonetheless). The new emulsions suspended solid grains of the silver salt in a thick fluid base. Egg white was used initially, then Frederick Scott Archer developed collodion, made from the explosive gun cotton soaked in alcohol, ether and potassium iodide. The mixture was spread onto a glass plate and sensitised with silver nitrate solution before being exposed. The collodion process improved sensitivity beyond that of Talbot's calotype method and it was cheaper than the daguerrotype, so it proved very popular. However, it was complicated and 'wet' – it used a variety of chemicals and required the photographic plates to be prepared and used before the collodion dried.

Astronomers were quick to take up the new technologies. French and American astronomers had been using the daguerrotype method to photograph the Sun and the Moon since the method was first introduced in 1839. In fact, the first picture of a solar eclipse was taken in 1851 at Königsberg, Prussia (now Kaliningrad, Russia), by an unknown local photographer, Berkowski, under the eye of the observatory director, Dr August Busch. In this early photograph, one can see the glow of the lower corona and several solar prominences. When the more sensitive collodion process was published it replaced the daguerrotype for astronomy, despite its inconveniences, and the first collodion picture of the Moon was taken in 1852. However, photography was still too immature for the aurora.

It was the introduction of mass-produced 'dry' plates in the late 1870s to early 1880s that revolutionised auroral photography. The idea came originally from Richard Leach Maddox, who in 1871 used dry gelatine as a base for the light-sensitive chemicals instead of a wet glass plate. This was not only much less messy, it was also transportable because the plates didn't need to be developed directly after exposure. In 1879, George Eastman, founder of Eastman-Kodak, invented a machine to commercially produce these 'dry plates'. Photography was becoming cheap and convenient. Just a few years later, during the first International Polar Year, in 1882–3, the Danish schoolteacher and researcher Sophus Tromholt made much use of the new dry plates when he was posted to Kautokeino in northern Norway.

The International Polar Year was a series of coordinated expeditions to polar regions around the globe, dedicated to scientific research. The idea was to obtain synchronous observations of 'the forces of nature in the Arctic regions.' The Polar Year was the brainchild of the Austrian explorer Carl Weyprecht, scientist and co-leader of the Austro-Hungarian Polar Expedition to search for a Northeast Passage in 1872–74 (they discovered Franz Josef Land, an island archipelago at 80°N, north-east of Svalbard and now part of Russia). He realised afterwards that though they had taken a multitude of measurements, they had no others with which to make comparison, so their data was of limited use. He proposed collaborative international expeditions, the first of which took place in 1882–3 across fourteen research stations – twelve spread around the Arctic – and employing around seven hundred men from eleven nations braving the conditions.

Tromholt had been working in Bergen as a teacher before he was stationed in Kautokeino for the first Polar Year. He established a small satellite station of the main Bossekop observatory 100 kilometres (60 miles) away so that synchronous measurements could be made with the intention of deducing auroral height. Whilst he was there, besides making accurate observations of the northern lights and venturing good explanations for observed effects, Tromholt took hundreds of pictures of the Sami people amongst whom he was living. These now constitute an important record of the Kautokeino people at that time and they somewhat obscured his work on the aurora, which was published as part of a travelogue entitled *Under the Rays of the Aurora Borealis*. Robert H. Eather, in his book *Majestic Lights*, describes Tromholt as 'the most underestimated auroral scientist of all time.' He effectively discovered the auroral oval, noted its movement over twenty-four hours and its expansion with solar conditions. He also attempted to photograph the aurora, but despite having the new photographic plates and exposures of several minutes he could not produce an image. It was a few years later, in Christiania (Oslo) in 1885, that he achieved a faint blurry auroral photograph with a long exposure of eight and a half minutes. This length of exposure blurred out all the detail and Tromholt never published the image. The first real photographs of an aurora, showing structure, were made by German meteorologist Otto Baschin and astronomer/photographer Martin Brendel in Bossekop in 1892, using highly sensitive plates and a wet process.

Thus as the twentieth century dawned, Birkeland was composing his theory of the northern lights and planning his terrella experiments as people were just beginning to photograph the phenomenon. Birkeland's enthusiasm inspired a young mathematics professor at the university, Carl Størmer, into a lifelong interest in the aurora. He began by studying mathematically the motion of charged particles in the Earth's magnetic field but quickly became an avid photographer, too. He took over 40,000 auroral photographs from twenty different stations around Norway in the years from 1910 until the mid-1940s. Many of these were paired observations, so that by using triangulation he could calculate the height of the aurora. He was the first to do so accurately. It was clear that most auroral light originated from around 100 kilometres (60 miles) altitude but there was yet a wide range in the heights. Størmer had collected so much data that despite publishing numerous journal papers and a book in 1955 there was still further statistical study to be done. Using Størmer's data, other scientists such as Alv Egeland and Anders Omholt were able to work with the data in subsequent decades, analysing the variation in height of different auroral features with magnetic latitude.

Talking to Eric at the University of Calgary, I was struck by what a tremendous achievement this was. The likes of Tromholt, Birkeland and Størmer pieced together the auroral oval by taking photographs well before the space age. 'They had the scales right,' said Eric. 'It was well above the surface of the Earth, it was above the atmosphere, and in a ring around the pole.'

It is much easier for scientists nowadays, with satellites and modern equipment. Størmer would have had to go out onto a windy mountain top, perhaps several days' hike away, and have a colleague go out onto another. They would synchronise the taking of pictures, trying to minimise errors in timekeeping, hoping that the cameras were trained on the same auroral feature. Now we have cameras that can automatically take a photograph, say, every six seconds, and we can set them up in multiple locations. The timings are set accurately by GPS*, so we know the cameras are taking pictures simultaneously. Scientists can easily access the data and compare observations from different locations at precisely the same time. They don't have to leave their offices.

The advent of the space age brought a new view, confirming some theories and refuting others. One beautiful confirmation was that of the auroral oval, and the Canadians had a hand in this too – in imaging on the large-scale. The Canadian satellite Alouette I, launched in 1962, was the first satellite to be built by a country other than the United States or Russia. It was designed in consultation with NASA for the purpose of mapping the ionosphere from

---

* GPS is the American Global Positioning System. There are other satellite navigation systems either in use or in development around the world, collectively referred to as global navigation satellite systems or GNSS, but GPS is the original and best-utilised of the world's satellite navigation systems. The term GPS may be used as a colloquialism to refer to GNSS.

above. For the burgeoning satellite industry, this would enable better communication from satellites to ground stations. Alouette I was one of the most complex satellites of its day and, in fact, NASA didn't think it would work and so didn't really plan to use the data. Also, in those days most satellites transmitted data back to Earth for a few months only, until their batteries ran out. Alouette I made use of emerging technologies such as transistors and solar cells, and it operated for ten years before being switched off.

Following the success of Alouette I, Canada and the US collaborated to launch three more satellites over the next decade under a programme called International Satellites for Ionospheric Studies. The satellites were called Alouette II, ISIS I and ISIS II and they were designed to study the ionosphere for communication and scientific purposes, including learning more about the aurora borealis. The ISIS II satellite, more advanced than Alouette in things like navigation and data collection, also included optical sensors made by the Auroral Imaging Group at the University of Calgary (the group that Eric now heads). These sensors were photometers, which measure the intensity of light at specific wavelengths. The photometers on ISIS II measured the three prime auroral colours of green, red and blue. Using these light measurements taken in strips over successive orbits of the satellite over the poles, the data could be processed to give images of the auroral oval from above. Merely kilobytes of data made up each image from ISIS II, and data for eighteen images was collected each day, more than could be processed with the computers of the time. The images are pixellated, indistinct, amorphous rings on a black background, but for the first time, in 1971, they confirmed the actual existence of the auroral oval.

NASA's Dynamics Explorer 1 satellite improved on the ISIS work, taking the first ultraviolet images of the entire oval from space in 1981 (although we only see the visible part, light emissions of the aurora extend out into the invisible parts of the electromagnetic spectrum and are also detected in the infrared, ultraviolet and X-ray). Measuring intensity in ultraviolet meant that the aurora could be detected even when visible light was obscured by sunlight, so the oval could be seen even as it stretched into daytime. 'Before the space age,' said Eric, 'they didn't know whether it [the auroral oval] was always there, if it is continuous. Now we know it's certain. If it's crystal clear and I get on a plane and fly from here [Calgary] to Resolute Bay, I know I will see the aurora.'

Dynamics Explorer was in a high-altitude orbit that stretched out to around 22,000 kilometres (13,670 miles), sufficiently high to see the whole auroral oval for two to three hours during a single orbit. It could collect enough data to produce an image every twelve minutes. Finally, scientists could track global changes in the aurora over time and put their more accurate ground-based images and measurements into the proper context. The satellites were beginning to get a glimpse into the processes going on in space to produce the aurora.

As I looked at the old pixellated pictures of the auroral oval I thought about how our ability to go into space had given us a whole new perspective, but also how that imprint that we see in our atmosphere is just the very end of the process, like that on the light-modified chemicals on the old photographic plates. As the Earth is buffeted by the solar wind our magnetosphere takes the blows, dissipating the energy in a reproducible cycle into our atmosphere – like breathing in and breathing out – and creating the familiar patterns of the aurora. I remembered something that Steve Milan had told me at the University of Leicester, back home in the UK. 'I'm a magnetospheric physicist,' he said, 'not an auroral physicist. I'm interested in what auroras tell us about how the magnetosphere is working.' There is much more to these pictures than initially meets the eye. Steve Milan said that how the magnetosphere works and how the aurora is generated is all to do with the interaction of plasmas. It's an intricate cycle but the auroral oval is always there, to varying degrees of strength, acting as a window to the activity in space beyond our atmosphere.

* * *

THESE INTERACTING PLASMAS that Steve Milan was talking about are where interplanetary space meets the Earth. Plasma from the solar wind meets plasma from the ionosphere, the part of our atmosphere above about 100 kilometres (60 miles) which is ionised by ultraviolet light from the Sun. Plasmas tend to have a magnetic field associated with them, which makes their interaction far from simple.

Plasmas with different magnetic fields can't mix – the magnetic field stops them intermingling. Imagine two ordinary gases blown towards each other; they would mix, like a scent gently wafting through a room. Not so for two plasmas; they will butt up against each other because the magnetic fields cannot cross. In space this creates boundaries and regions occupied by plasmas from different sources. The Earth sits in its own little protective bubble of plasma within the stormy flow of the solar wind.

The Earth has a dipolar magnetic field – the geomagnetic field – which creates the butterfly-wing magnetic field pattern of a bar magnet around it. The solar wind is travelling out from the Sun into space, pervading the solar system in all directions, and the solar wind has its own magnetic field frozen into the flow of moving plasma particles. When the solar wind hits Earth it is deflected around it, like water in a river rushing around a stone. This reshapes the Earth's magnetic field – squashing it up on the Sun side and stretching it out like a windsock behind the planet. This creates what we call the Earth's *magnetosphere*, the region in space where the influence of the Earth's magnetic field can be felt. It is not symmetrical, rather a sideways teardrop shape with its bottom towards the Sun, Earth glinting in the middle of the widest point.

The magnetosphere acts as the Earth's shield, protecting it from the influx of the fast-charged particles of the solar wind. Without the magnetosphere Earth may not have an atmosphere, and instead could be a barren wasteland as on Mars, whose primordial atmosphere was stripped away by the solar wind over billions of years. This high-energy radiation – the fast solar wind particles – is also a threat to life on Earth, capable of damaging DNA molecules and cells. The magnetosphere and the atmosphere together make Earth habitable.

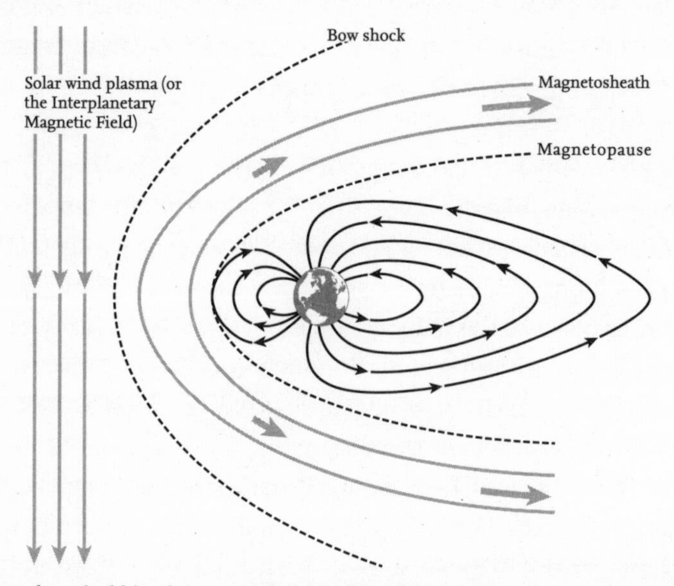

As the solar wind hits the magnetic field of Earth it is deflected around the outside, flowing in a layer called the magnetosheath. The Earth's magnetic field is stretched out behind. The cavity of the Earth in the solar wind is the magnetosphere and its boundary is the magnetopause.

The particles deflected around the edge of the magnetosphere, called the *magnetopause*, flow in a region called the *magnetosheath*, bounded on the Sun side by a shock front where the solar wind is slowed from supersonic to subsonic speeds. This idea of the magnetosphere was proposed in 1930 by the English physicist Sidney Chapman and his student, V.C.A. Ferraro. They realised that this solar wind cloud of ions and electrons (the newly coined name of 'plasma' had not yet stuck) would be a good conductor of electricity, so electric currents would be induced to flow in it as the cloud approached the Earth's magnetic field. These currents that flow on the magnetopause surface contain the Earth's magnetic field and so define the magnetosphere, which for many years was known as the Chapman-Ferraro Cavity.

The magnetosphere is often described as a cavity embedded in the solar wind, like a bubble. The region of space occupied by a magnetosphere is much less dense than the surrounding solar wind, except at distances very close to the planet where the atmosphere bleeds out into space. The solar wind at Earth's orbit contains about ten particles per cubic centimetre, whereas some of the denser parts of the Earth's magnetosphere contain about one particle per cubic centimetre, and there are huge volumes of the magnetosphere where the density is even less. Compare this with water, which has about thirty thousand billion billion molecules per cubic centimetre. Within the magnetosphere, there is hardly anything there; it is the magnetic field of the planet that is providing the pressure to keep the solar wind out, not the matter itself. So rather than thinking of the magnetosphere as a rock in the river of the solar wind, we could think of it as a bubble stationary in the flow.

The size of the magnetosphere depends on the buffeting it is getting from the solar wind. If you squash a plasma you increase the field strength inside and the plasma pressure resists being squashed. It reaches the point where it can't be squashed any more. This is why the magnetosphere deflects solar wind.

The pressure of the solar wind hitting the magnetosphere is balanced by the magnetic pressure exerted by the geomagnetic field, which increases as it gets more squashed up. For typical solar wind conditions the edge of the magnetosphere on the Sun side is at a distance of about five times the diameter of the Earth, but it will be pushed in closer if the solar wind is strong. The tail side can be stretched out, typically to a distance of five hundred Earths.

Despite the protective shield of the magnetosphere, particles do get in and out. When Birkeland was doing his experiments with the terrella in the early years of the twentieth century, setting up the magnetised brass sphere in the path of cathode rays so that its poles glowed bright with an artificial aurora, he was convinced that the northern lights were caused by electrons from the Sun guided by magnetic field lines to the polar regions. The electron had only been discovered by the English physicist J.J. Thomson in 1897, who determined that cathode rays were streams of these tiny negative particles. The situation with the aurora was, in fact, not as simple as particles from the Sun simply getting caught in the magnetic field of the Earth and channelled to the poles, but it would be several decades before the real picture was seen.

In the early nineteenth century it was known that big magnetic storms on Earth were connected with sunspot activity. Birkeland believed that his solar electrons would be emitted by sunspots; however, at that time basic knowledge of the Sun was insubstantial. The continuous solar wind was not known then, nor even the mechanism by which the Sun shone. Radioactivity had just been discovered and the idea was put forward that radioactive decay could produce the energy required to accelerate these solar cathode rays. Radioactive decay was not powering the Sun, but the real process of nuclear fusion was still to be found. With such a limited understanding of the workings of the Sun, scientists advocating theories involving cathode rays coming from the Sun generally thought of those particles arriving at Earth in bursts.

Despite his best efforts, Birkeland's theory of the aurora was not recognised by the scientific establishment in his lifetime. It was criticised by the German-born, British-raised physicist Sir Arthur Schuster, a Fellow of the Royal Society, who said that a beam of similarly charged electrons could not hold together under electrostatic repulsion. Birkeland answered this criticism in Part II of his book *The Norwegian Aurora Polaris Expedition 1902–1903*, published in 1913, and indicated in a 1916 publication that he believed there to be both negative and positive 'rays'. However, this slighting by a prominent British scientist, later perpetuated by Chapman, did not help the acceptance of Birkeland's theory.

Surprisingly, Chapman himself, as a young man in 1918, had proposed negative particle streams as causing magnetic storms. It was Frederick Alexander Lindemann, the Oxford professor who would later become the Second World War advisor to Winston Churchill, who quickly disputed this hypothesis for the same reason that Schuster had dismissed Birkeland's. However, Lindemann made the suggestion that the charged particle stream was electrically neutral, composed of a mixture of positive and negative charges (protons and electrons).

Throughout the early and middle years of the twentieth century, scientists such as Chapman formulated their theories of the solar particle streams and the magnetosphere while the solar physicists puzzled over the workings of the Sun and spectroscopists studied the auroral colours. It was 1958 before the concept of the solar wind was introduced. The American physicist Gene Parker suggested that the charged solar particles didn't come in bursts or clouds but as a continuous stream, the atmosphere of the Sun being so hot that it was expanding out into space. Concurrently, in the 1950s, Parker was one of the main protagonists in forming the theory of how the shield of our magnetosphere is penetrated by solar particles that go on to cause the aurora.

It is true that some particles can enter directly into the Earth's atmosphere. There are holes in our shield: two areas of weakness at the poles where the Earth's magnetic field opens out into space like a funnel and charged particles directly from the Sun can enter the Earth's atmosphere. These are called the polar cusps. However, this direct funnelling comes from the Sun-side of Earth, which results in daytime aurorae – an effect that cannot be seen in daylight. Day-side aurorae are only visible during polar night when it is dark at midday. Even then they are less bright than night aurorae and usually red in colour, which to the eye is less sensitive than green, so it was only discovered in the 1960s that the aurora occurs during the daytime too. However, although the day-side aurora is interesting scientifically and tells us immediately about solar wind conditions, it is not the true night-time aurora with which we are all familiar. An entirely different process in the magnetosphere is responsible, a cycle that takes the incoming energy of the battering solar wind and dissipates it in an atmospheric display of great beauty. The secret of this cycle is the mysterious magnetic reconnection, the key that opens up our magnetospheric shield to the solar wind.

* * *

THE MAGNETIC FIELD direction in the solar wind can be pointing north or south, and it doesn't always hit the Earth head on, for it is often tilted. The Sun is that spinning ballerina and her skirt spirals out, marking the hard division of North and South polarities of the magnetic field of the spiralling plasma. As the skirt rises and falls, the field direction of the solar wind heading towards Earth varies.

When the solar wind hits the magnetosphere the plasma is diverted around the Earth. Since the magnetic field is frozen into the plasma the magnetic field ends up draped around the magnetosphere. This leads to a clustering of magnetic field lines just upstream of the Earth, towards the Sun, 60,000 kilometres (37,500 miles) above the equator. Here the magnetic field is intense and the crush squeezes plasma particles out of this region along the field lines. But this is not all that happens in the crush; if the magnetic field changes a lot over a small distance, for example in the case where two fields pointing in different directions are pushed extremely close together, then the frozen-in condition of magnetic fields in plasmas can break down and the magnetic configuration can change. This is a process known as *magnetic reconnection*, first proposed by British physicist Peter Sweet in 1956 and developed mathematically by Gene Parker. If two plasmas with oppositely directed magnetic fields meet but cannot cross, they begin pushing closer and closer together and the magnetic field lines become more and more tightly packed, down-up lines pushing against up-down. Eventually the magnetic fields will be so intense in such a small spatial region that the field lines will break and reconnect one by one, the first plasma's down field connecting with the second plasma's up and twanging out sideways as the magnetic tension is broken. As well as magnetic fields having a pressure associated with them, they also have a tension, which means that they act a bit like rubber bands. If the field is highly bent it will try to straighten up as quickly as possible, catapulting plasma away from the reconnection region and down the field lines. In this way, energy is transferred from the magnetic field to the particles, which are accelerated to high speeds. Aurorae occur at the footprint of the magnetic field lines.

Magnetic reconnection is a fundamental plasma process that occurs in many situations throughout the Universe. Coronal mass ejections from the Sun are governed by reconnection, and the phenomenon is becoming of increasing interest to the astrophysical community to explain some astrophysical processes. When reconnection happens between magnetic fields from the Sun and the Earth, it starts off the magnetospheric processes that cause the aurora. Reconnection is a fundamental part of the picture, and yet we still don't fully understand it.

The most vivid aurorae on Earth occur when the magnetic field of the solar wind is directed southwards. In this case, it points in the

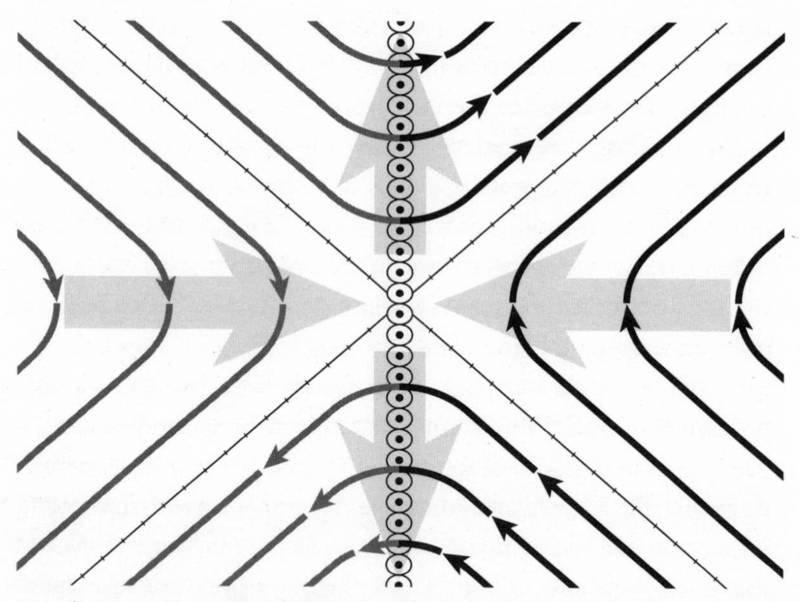

Two plasmas come together and their oppositely directed magnetic fields get pressed too close. Eventually the field lines will reconnect and plasma will be catapulted out of the region as the newly joined field lines straighten out.

opposite direction to the Earth's magnetic field, which points from the geographic South Pole to the North Pole*, that is northwards. Since the fields are oppositely directed, magnetic reconnection can occur. When the solar wind impacts the magnetosphere and reconnection occurs between Sun and Earth, there is now an opening in the magnetosphere on the Sun-side at the level of the equator. The magnetic field lines open out and are dragged around the Earth away from the Sun, being pulled into the tail region. Field lines that were once 'closed' – looping from pole to pole on Earth and confined only to the magnetosphere – are now 'open' and they

* Magnetic fields are directed from a North Magnetic Pole to a South Magnetic Pole, but since the geomagnetic field is currently oriented with its North (Magnetic) Pole in the south (geographically) and its south in the north, the geomagnetic field points geographically south to north.

connect out into the solar wind. The shield has been broken and solar wind plasma particles can get in, following the magnetic field lines into the magnetosphere.

All the time that the solar wind's magnetic field is southward, new plasma is coming in towards Earth and opening up field lines, pushing them around into the tail so that more and more plasma is being laid onto the top and bottom of the tail region. This pushes the sides of the tail in towards each other, and once again oppositely directed magnetic field lines come into contact.

When enough new tail plasma is pushing down, magnetic reconnection will occur in the centre of the tail. This changes two open field lines – hybrids of the magnetosphere and solar wind fields – back into two closed field lines, one connected only to the Earth and the other in the solar wind. In this way the solar wind has passed around the Earth, like that stone in a stream, and the solar wind flow has rejoined behind the Earth. At the same time, the energy of the solar wind hitting the Earth has been absorbed by the magnetosphere and dissipated in the acceleration of plasma particles.

The reconnection process creates new, highly bent field lines that instantaneously straighten up. The newly closed field lines contract back towards Earth, catapulting particles down towards our atmosphere. This forms very vivid aurora on the night side of the planet. We picture this process in two dimensions to aid our understanding, but of course this is a three-dimensional system. The reconnection region in the tail is not simply a point but a line of reconnection stretched out behind our planet. After reconnection, the newly closed field lines push back towards the day side, drifting sideways round the Earth. This is called the Dungey Cycle, named after Jim Dungey of Imperial College London, who originally proposed the idea in the early 1960s. It is the cycle of the field lines, and of all the plasma within the magnetosphere constrained to follow them.

As this process plays out, from day-side reconnection opening field lines to night-side tail reconnection closing them, the magnetic topology of the magnetosphere is changed. Looking down on the Earth from above, if we could see the field lines we would see a central hole of open field lines stretching out into space around the poles. As we moved further from the poles we would see them change to closed field lines, bending back down towards the opposite pole. In fact, though we can't see the field lines themselves (remember they are not real, but a visualisation construct invented by Faraday), we can see the transition from open to closed. It is at the boundary where open field lines reconnect into closed ones that we see the aurora, like the footprint at the end of the field line. Seen from space, this footprint takes the shape of an oval of light around the pole – the auroral oval – and the dim region inside the oval is where the open field lines stretch out into space.

The most impressive aurora on Earth are due to the bursts of reconnection in the tail when the solar wind magnetic field is southwards. It occurs explosively and fires particles down towards Earth on the night side, which is why the best and brightest aurorae – even at winter solstice when the day side is dark – are seen on the back side of the Earth. These explosive events are called substorms.

Reconnection is triggered through a sequence of events that has been the subject of study for at least forty years and that no one really understands properly yet. The Dungey Cycle offers a good global picture, as described, but the actual onset of reconnection has to do with the microphysics that is going on in the reconnection regions. Currents are also involved because they exist due to the movements of charged particles and at the boundaries of plasmas, such as around the edge of the magnetosphere or bisecting the tail. The whole system is interconnected. Around reconnection events the tail current becomes disrupted and some current flows along the field lines into the ionosphere. No one knows yet whether the disrupted currents trigger the reconnection

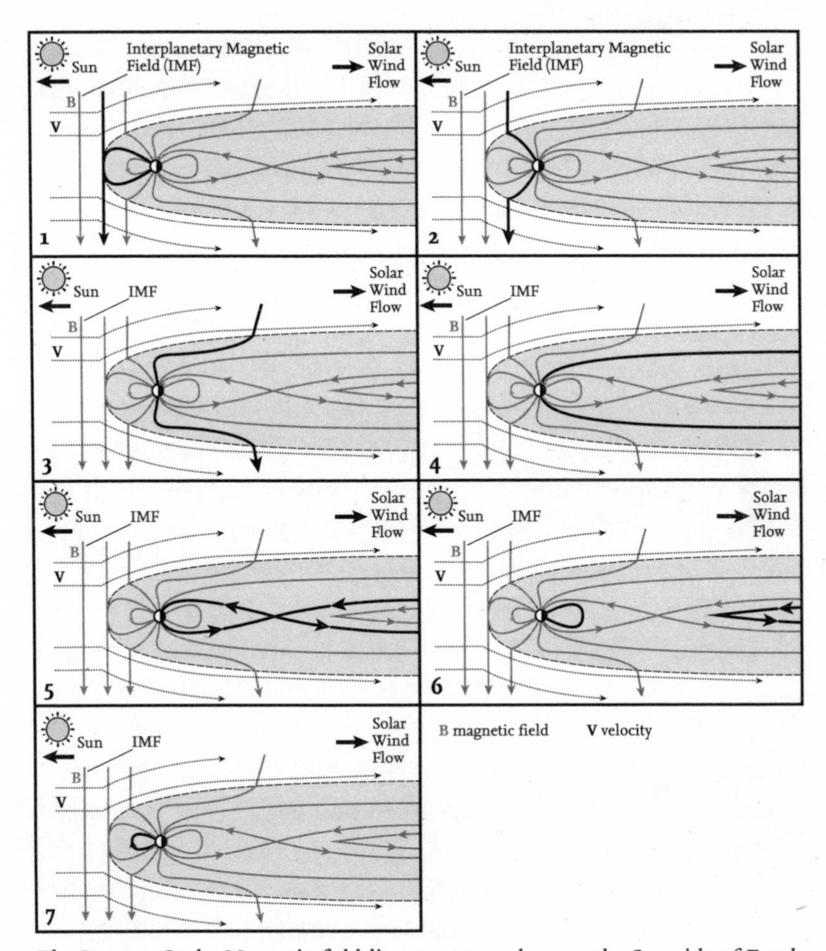

The Dungey Cycle. Magnetic field lines are opened up on the Sun side of Earth, dragged (with their plasma) over the poles and laid down on the tail side. Here they push down until reconnection closes the field lines once more and they migrate sideways back to the Sun side of the Earth. This cycle can continue as long as the magnetic field of the solar wind is pointing southwards.

or whether the reconnection triggers the currents. This is one of the biggest open questions in the field and is known as the 'substorm problem'.

Many of the cameras dotted around Canada, like the ones I saw in Athabasca, were installed as part of the THEMIS mission, which was trying to solve this substorm problem and understand the trigger mechanism. Scientists wanted to know what it is that causes the aurora to suddenly become much brighter and more dynamic rather than just appearing as a green ribbon stretched across the sky. To do this they had to identify where the substorm originates. What finally snaps the over-stretched magnetic field lines in the magnetotail?

The two possible triggers – currents or reconnection – happen in very different parts of the magnetosphere and it's difficult to tell which occurs first. The currents that flow across the tail region are about 60,000 kilometres (37,000 miles) from Earth, but where the ends of the tail get squeezed together and reconnect is about 120,000 kilometres (75,000 miles) from Earth. If the current across the tail drops away, it could cause a wave to ripple outwards through the tail and instigate reconnection, but on the other hand reconnection would cause particle flows to disrupt the current. Whether it's an inside-out or outside-in process is a very big question. 'So it's perfect for a NASA mission!' declares Eric.

NASA's THEMIS mission, launched in February 2007, attempted to solve this mystery by combining a fleet of spacecraft with extensive ground-based imaging. The name is symbolic; as well as being a scientific acronym (as these things often are) for Time History of Events and Macroscale Interactions during Substorms, Themis is the name of the Greek goddess of order and justice who, in NASA's words, 'will resolve this debate like a fair, impartial judge.'

THEMIS used five identical satellites which would line up every four days over North America, stretching out downstream of Earth into the magnetotail. All-sky cameras in 20 ground stations across Canada and Alaska imaged the aurora from below, while the satellites monitored conditions in the magnetosphere above.

The mission collected data for two years and it is still being analysed today. Since then, three spacecraft have continued to study substorms while two were sent on a new mission to the Moon, to observe the effects of the solar radiation on a body without the protection of a magnetic field.

Eric's team at the University of Calgary is responsible for the network of cameras in Canada. They cost $1–2 million to build and deploy out of a total mission cost of around $200 million, so they were only a small part of the budget, yet vitally important.

'The cameras are as important as the in-space measurements,' Eric told me over lunch one day. 'It's much more than just context. The cameras show how the aurora is unfolding in time and we are using this to explain the physics.' It is possible to do this because the auroral arc maps to a region out in the magnetosphere. Eric thinks that because of THEMIS the data now exists that can answer this big question and solve the substorm problem. In doing so, the cameras across Canada have been vital. 'The information we need on this scale size,' he said, 'is best captured by the aurora and the all-sky imager data.'

None of this happens if the incoming magnetic field in the solar wind is directed northwards. When the solar wind magnetic field points north the pattern is different and the aurora is weak. Then the solar wind really does flow around our planet as if it were a stone in a river. The reconnection site will be at higher latitudes where it butts up against Earth's magnetic field directed south. Often, in this case, the solar wind field reconnects with field lines that are already open – as opposed to when the field is southwards and reconnection opens up closed field lines – so there is no real change in the magnetic pattern. This results in flows in the polar region but no widening of the auroral oval. Weak, predominantly red, aurorae may be seen at very high latitudes, but little else. It is only when the magnetic fields pressing against each other are oppositely directed that there is an option for magnetic reconfiguration, a place where the fields can snip and rejoin and still be following the same direction as before, but in a new pattern and having transferred some of their magnetic energy to the plasma particles.

So, with southward field, the cycle continues. Field lines reconnect and open on the Sun side, are dragged up and over the poles, pushing down on the tail where they reconnect again and close, springing back towards Earth and moving back around the equator to the Sun side again, where the process can repeat. A very eminent space scientist from NASA's Goddard Space Flight Center once described this to Steve Milan as Magnetospheric Tai Chi: open; over; close; around; repeat.

CHAPTER FIVE

CANADA – COLOURFUL COLLISIONS

After Athabasca, I left Martin in Edmonton and flew further north to Yellowknife. I saw it first from the air through a smokey haze; forest fires were raging in the Northwest Territories, one as close as 30 kilometres (19 miles) outside of town, in their worst wildfire season in 30 years. Yellowknife spread out between the irregular patches of sparkling water, clusters of houses and tower blocks in a land of lakes, rock and trees. It was hot and bright, and as I disembarked from the plane the air smelt of campfires – that distinctive smell of burnt wood.

After leaving my luggage at my hotel, I walked to the old town, down the hill and across a short spit to where houses were built up and around a large, rocky outcrop. I had a salad on the wooden terrace of the Wildcat Cafe, the oldest eatery in town with a tumultuous history of ownership, closures, re-openings and renovations since it first began serving in the 1930s. An article in a 1938 issue of *LIFE* magazine called it the local 'hotspot'. The vintage log-cabin building is now a protected heritage site, a vestige of the original mining community in Yellowknife. I sat and watched the float planes coming in over the lake.

On a map, the Great Slave Lake (the fifth largest in North

America) looks a little like a large-headed goose in flight, halfway from the Alaskan border to Hudson Bay. Yellowknife sits atop the goose's wing, a small city on the lake shore and the capital of the Canadian Northwest Territories. Great Slave Lake is part of the giant Mackenzie River waterway, connecting south to Athabasca, where I had been just a couple of days before, and flowing north-west for a thousand twisting miles through a vast, island-studded delta to drain into the Beaufort Sea in the Arctic Ocean. The lake is frozen for eight months of the year and in winter forms part of the ice road linking Yellowknife to other regional communities.

Covered in snow and ice in winter, northern Canada in summer sees water everywhere. The big lakes like Great Bear and Great Slave are prominent, but the whole region moving eastwards from the Rocky Mountains is peppered with lakes of various sizes. From the air it is as if glitter had been sprinkled in a diagonal stripe across the northern reaches of the continent.

Since the Europeans began exploring the Arctic in the search for the Northwest Passage, men have been prospecting, searching the land for valuable mineral deposits to exploit. Many of the northern towns started up in prospecting rushes for minerals such as gold, copper, uranium, lead, silver or zinc. Consequently, the towns would boom and slump. Even the indigenous peoples, before the arrival of the Europeans, were using copper. Yellowknife is named not for the gold, but in honour of the local Dene people who carried knives with copper blades.

Yellowknife had a relatively slow start in mining terms. There had been activity around the Great Bear Lake, further north astride the Arctic Circle, when in September 1934 some prospectors paddled south. They followed a lake-and-river trail down into the Great Slave Lake, where a storm drove them ashore. Here the party found gold and ignited a small gold rush. A tent city sprung up on Yellowknife Bay, land was staked which would later become Yellowknife's largest two mines, but activity slowed within a year. It wasn't until 1936 that an Englishman, Tom Payne, beached his birchbark canoe not far from the mining camp and, whilst idly panning in the evening, found good gold there. Unfortunately for him, the land had already been claimed by a Toronto prospector who was not yet exploiting it. Payne waited three months in secret for the prior claim to lapse before bounding out in torchlight at midnight on 27th September to stake out his land claim. This land was adjacent to one of the two largest mining areas owned by the Consolidated Mining and Smelting Company – and they wanted it. In an unprecedented negotiation, Payne's representative secured half a million dollars for a 60 per cent share in Payne's mine and the Yellowknife gold rush began.

Yellowknife became the fastest-growing town in Canada in the biggest northern gold rush since the Klondike. Yet this time it was different. Now prospectors could fly in by float plane rather than mushing in by dogsled, and the town boasted a number of modern conveniences. The Con Mine, the Northwest Territories' first gold-mine, opened in 1938 and operated for sixty-five years before closing in 2003. Most of the ore around Yellowknife is low-grade and so required large-scale mining to extract it profitably. The gold-mining boom was helped along by currency devaluation during the Great Depression in the 1930s, which increased the price of gold and made mining in more inaccessible and less richly endowed areas profitable.

Mining was suspended or slowed during the Second World War but quickly restarted afterwards. The second of Yellowknife's main mines, the Giant Mine, opened in 1945 and the town expanded to its new site in 1947, exchanging the log cabins and tents for grid streets and larger buildings. The government established in the 1930s developed into a fully elected Legislative Assembly by 1975, with Yellowknife becoming the capital of the Northwest Territories.

By the turn of this century, diamond mining had replaced gold mining in Yellowknife. The diamond mines were further from town but accessible from Yellowknife. Industries grew up around them; oil and gas exploration now also contributes to the town's prosperity.

Meanwhile, the mines that made Yellowknife are today causing the town trouble. Long after the original prospectors' millions have been made, billion-dollar clean-up plans are being funded by the public purse since the mines closed. Highly toxic arsenic-rich waste lies in ponds around the mine sites, the legacy of the intensive mining process.

Nowadays, as the only city in the Northwest Territories – and the capital – Yellowknife is a communications and transportation hub. Trucking is a major industry, with ice roads fanning out from Yellowknife to all the local mines. In the 1950s, a tall, lean, ex-Mounted Policeman with a mechanical mind and northern ingenuity pioneered the ice roads. John Denison believed it was possible to build a winter road strong enough to support convoys of trucks and improve vital connections to the mines – which would be cheaper and safer than flying and faster than taking massive snow-cats that crawled at 8 kilometres (5 miles) per hour. By ploughing away insulating snow and exposing the ice to the frigid air, they would thicken and strengthen it, later dragging tyres over it to even out the surface for the convoys. Even now the ice roads draw the best truckers, those with the road skills necessary to drive on the ice, and the wherewithal to take the challenge and the hostility for the double pay and the adventure. In those early years, Denison and his team endured foul weather, rugged terrain and isolation to build over 1600 kilometres (1000 miles) of routes and get the cargo through to the mines. They had to deal with whatever conditions or eventualities came their way, from breakdowns and failures to trucks falling through the ice into lakes and rivers. And each season they would do it all again.

For many decades, going right back almost to the beginnings of the ice roads, there was talk about building a permanent, paved road beside the Mackenzie River. This would replace the Sahtu winter road to the oil and gas explorations to the west of Great Bear Lake at the Arctic Circle. In the last decade, traffic has been drastically increasing in these areas – sometimes one to two hundred 40-tonne trucks per day drive on a road of snow and ice that ruts under the heavy loads. Companies have only a three-month window while the ice road is open to transport cargo to their operations up in the North, unless they are prepared to take the slower option of sailing it on barges up the river in summer. Good planning is key, but so are good roads that can take the strain. Operators in the region believe that an all-weather road is an important step towards full-scale development and that the government should support the exploitation of its assets. It would also bring cheaper prices and year-round access to remote communities along the route, but of course there are worries about what else it would bring – the alcohol, drugs and unsavoury characters.

The Government of the Northwest Territories says it is now committed to the design, development, construction and maintenance of what is referred to as the Mackenzie Valley Highway, a road linking Highway 1 from Wrigley, on the Mackenzie River west by north-west of Yellowknife, through the Mackenzie Valley to the Arctic Coast. There is a desire to further exploit the resources of the Canadian North – renewable as well as non-renewable. Rather than an inaccessible backwater, the government is now looking on the North as an opportunity. In a global marketplace it needs to help northern businesses compete on a global scale. It also sees the North as a gateway; as climate change grips and Arctic sea ice melts, the prospect of an ice-free, navigable Northwest Passage – the old dream of early explorers – presents the Canadian polar regions with new possibilities.

As the seat of government in the Northwest Territories, the Legislative Assembly in Yellowknife brings together nineteen representatives from small, disparate communities, balancing modern development with indigenous heritage. The first Inuit member was appointed to the Northwest Territories Council (a decade later to become the Assembly) in 1965, two years before Yellowknife was named as capital. Representatives from other groups such as the Dene soon followed. Aboriginals are involved, elected and celebrated, yet it can be a challenge to capture the spirit and needs of the people without romanticising or trivialising the Arctic.

Whilst in Yellowknife I visited both the Prince of Wales Northern Heritage Centre and the Legislative Assembly to gain a meagre understanding of the aboriginal history and the cultural context. It can be hard to hold on to a lifestyle and a culture in the face of modernisation and globalisation, particularly for the minority cultures. It happens all over the world. Mike Mitchell at the Heritage Centre told me how many of the aboriginal languages are at risk of dying out. There are eleven official languages in the Northwest Territories, encompassing Dene and Inuit communities in addition to French and English. Language preservation is a huge issue, particularly in the more connected communities where often only the elders still speak the original tongue. Subsequent generations may understand the language but often don't acquire it fully. Other elements of culture are preserved but changed, as modern conveniences are absorbed into traditional lives.

I asked Mike what people thought about the proposal for better road transport. Do they think of it as progress and welcome the better connections, or as an unwelcome threat to the landscape?

'People have different opinions for sure,' said Mike. 'It seems to me that whenever there's access into a wilderness area then that's when you get pressure on the ecosystem.' He told me that the caribou herd in the region had declined from 350,000 to 35,000 in a little under two decades. It is quite likely that some of the decline can be attributed to the diamond mines and ice roads that pushed up into their winter range. 'It's a tricky dynamic,' Mike conceded. 'There are environmental NGOs in Yellowknife who think that we should be preserving the ecological value of these territories, but then if you're living in Tulita or Déline, which are connected only by ice road, and you're paying $8 for a litre of milk, then you probably want the road.'

'I suppose,' I said, 'that it's important to remember that people actually live there. The land is not just empty, there are communities, so there is a value to them to be more connected.' I remembered a story I was told separately about how the biggest single driver in the shift from traditional to modern was when television came to the North in the 1970s and 80s. As TVs started popping up in living rooms, suddenly young people were seeing things that they never knew existed before – Nike shoes, elaborate parties, TV shows, sporting events – and some of them wanted to be part of it. It wasn't a slow introduction to the outside world; by that time television was a developed system and indigenous peoples were given an instant view into consumerism and all that it entails. The voices of the elders were drowned out by the sound of the television.

Changes in lifestyle mean that people need the money economy and they need to work. The elders, who remember a different lifestyle, may lament the fact that things are changing, but we all have a tendency to do that anyway . . . in every country. Mike said that the impression he gets is that the elders in the Northwest Territories are generally accepting of newcomers to the region and the ethnic changes that creates; they are more concerned about preserving land and the animals for posterity. A fair concern, I think.

It is easy to view the Arctic as an empty, romantic region, but the critical thing is that the Arctic is not empty, and hasn't been for thousands of years. Surely we cannot deny those communities the privileges of modern life to preserve our romantic ideal, but neither should we exploit to the extent that we take away a livelihood founded on a balance with the land, or force that transition if it is unwelcome. The big question is how can we use the land, and perhaps even exploit the riches, in the Arctic in the kindest way possible? Human greed has a history of neglect, disrespect and flagrant carelessness. Whether we can expect anything better now, as we melt the icecaps and rush in to take the emerging spoils, remains to be seen.

The Legislative Assembly building, as well as being a functional parliament, is a monument to the diverse cultures of the Northwest Territories. 'One Land, Many Voices' is inscribed in the official languages on the new ceremonial mace, made when the once-giant Northwest Territories was cleaved in two in 1999 to create Nunavut as a separate province. As I toured around the Legislative Assembly in Yellowknife, examining the beautiful Arctic art and listening to the stories of the tour guide, I thought about a similar tour I made just a few months before. Looking round the Sami parliament building in Karasjok, Norway, we had heard of the similar issues of finding balance between indigenous values, national identity and development. The Sami herders in Scandinavia travel across the whole Finnmark plateau, traversing Norway, Sweden, Finland and even into Russia. Reindeer don't see national boundaries. I thought about Knut, the reindeer herder, and his fears for his herd and his children's future; of Cecilie and her family; of the fire and the smoke in my face as I sat on a reindeer skin in the snow.

Meanwhile, we had moved into the Chamber of the Legislative Assembly building, round in shape and topped with a shallow dome, designed to be reminiscent of an igloo and to represent both cultural traditions and consensus government. A polar bear rug sprawls in the centre, bearing its teeth to visitors. In the 1990s, one of the members of the Assembly from one of the high Arctic constituencies of what would soon become Nunavut approached the then Minister of the Environment and Natural Resources with an issue. The member told the minister that they were having a real problem with polar bears in their community. The bears were coming into the community, eating their food and posing a danger to people. The member said that the situation needed to be addressed promptly. The minister prevaricated, saying he had a lot of other things to do but that he would get to it soon. One day, before the minister did indeed get around to doing anything, the member walked out of his front door and a polar bear was standing right there in his driveway. Fortunately he had a rifle on him, so he shot the bear in self-defence. The member had the skin made into a rug and brought it back to Yellowknife as a gift for the Minister of the Environment and Natural Resources. It now lies in the centre of the Chamber as a reminder that one is never too busy to take care of the issues that the members address.

To better understand and improve the economics of the North, the government has to understand the socio-economic landscape as well. There is a lot to learn, and they are still figuring out the best approach. These days the government is the largest employer in Yellowknife, surpassing the mining trades and associated industries. Tourism is also on the rise and increasingly important to the local economy. The long summer days and the lake-shore location draw visitors for hiking, fishing, boating and to see the wildlife. Yellowknife is the Canadian city with the most hours of summer sunshine. Then in winter, of course, Yellowknife has the aurora, and one of the residents with the greatest appreciation for the nightly light show is James Pugsley.

* * *

I WOULD MEET James in the evening, after dinner at around 9pm, when it was still light and felt more like six or seven. We would drive out somewhere as we talked, James taking me to see the old gold mines on the edge of town with their 'danger' and 'open pit' signs, discarded foreign materials – a piece of long black tubing here, a rusting metal stump there – in a dune-like landscape of rocky, grey rubble grown over with small trees, bushes and purple fireweed* that slowly gave way to the solid rock of the Precambrian Canadian Shield. Or we would head out to a small lake surrounded by low trees, where in the winter, James said, the aurora would dance above them. The cirrus clouds and smoky air lent a mottled purplish-blue to the sky as I took a photograph of James, standing solidly in a wide stance, with arms folded across his blue-striped

* Rosebay willowherb to the British.

polo shirt, and smiling. He had a kind face with dark hair and a small, neat beard and wore thin glasses that I could barely see from where I stood. In the picture the lines of cirrus almost mirrored the lake's ripples and the water took on a faint pinkish hue. Or we would go to a Tim Hortons drive-through for coffee, despite it being 10pm, just because, as James said, it's a Canadian institution.

James Pugsley is an amateur astronomer, ex-journalist, government employee and president of Astronomy North, a volunteer organisation for education and outreach about the northern skies. The far northern, or southern, sky can be very different to that shown in textbooks based on lower latitudes. The most obvious difference is the view of the Sun. High latitudes see dramatic seasonal effects like long, sometimes endless, summer days and dark winters. The view of the Moon and the stars is also different. Around the shoulders of the solstice, a month or so either side, sky watchers may be treated to a display of noctilucent clouds – bright blue wisps of extremely high-altitude ice crystals reflecting light from the newly set sun into the twilight sky. Then from September to May there is the aurora, and James will be out with his cameras.

James runs the AuroraMAX project, a collaboration between Eric's team at the University of Calgary, the Canadian Space Agency, the City of Yellowknife and the Astronomy North Society. It is an outreach initiative that aims to show the splendour of the northern lights, raise awareness of the science and also showcase Canada's scientific interest in the aurora and some of the leading projects like THEMIS. James gathers data to show the intensity and frequency of the northern lights and to demonstrate that they are not just an occasional thing for Yellowknifers. The aurora is happening almost every night.

'We have consistently seen substorm activity above Yellowknife whether there's an active sun or not.' James told me. Yellowknife is in a prime spot and is often described as the best place in the world to see the aurora. It is the unique combination of perfect magnetic latitude and arid climate. It sits right in the auroral zone in the centre of a continent and in the rain shadow of the Mackenzie Mountains to the west. Consequently it has mostly clear nights, perfect for aurora viewing.

Initially the aim of AuroraMAX was to connect a Canadian audience with what was going on in their back yard, but a side benefit is that they showcase their aurora to the world, broadcasting the display above Yellowknife via the Internet every night from September to May. The AuroraMAX website gets more than 13 million hits over the course of the winter, and on the Canadian Space Agency site the page that gets the most hits by a long way is the AuroraMAX page. People just love the aurora.

One night, after watching some incredible 3D aurora videos in his flat, James took me to see the AuroraMAX observatory. We drove out of town and through the stunted trees until, just as it was beginning to get dark, we pulled up outside a compound with a high chicken-wire fence and gate. As soon as we got out of the car we were fighting off the mosquitoes. The heavy chain clinked as James unlocked the gate, telling me how the bugs didn't seem that bad and that usually he would have four or five mosquitoes biting his face while he was fumbling with the chain.

'And in the winter doing this is brutal,' James quipped to me. 'You can't touch the metal.' I envisioned trying to open the gate wearing heavy mitts, probably removing them and using only liner gloves to get the key into the lock, trying to work as quickly as possible to prevent fingers seizing up in pain. Tonight it was warm, but I slipped on my jumper anyway just to keep the bugs off me.

Once the gate was open we dashed for the protection of the observatory, our feet crunching on the gravel as we ran. The observatory was actually a wooden garden shed with a dome cut into the peaked roof – all you need for automated auroral imaging, which is controlled remotely by the team in Calgary. The left-hand corner of the shed housed a custom-built but ordinary-looking desktop computer specifically for the purpose of imaging. The dome housed the cameras, or would have done had they not been dismantled for the summer. At this time of year the sky never gets dark enough to see the aurora properly. The dome was shielded from the computer area by a thick black curtain to block out any stray light from blinking LEDs. Two dishes were fixed to the outside to link up to satellites for the Internet and to provide the live feed of aurora images. Despite not having wired Internet at the observatory, they did have electricity. A challenge for all the small observing stations that housed the ground-based THEMIS cameras was finding dark locations with power for the computers and cameras.

In the stuffy shed, James told me more about the project.

'AuroraMAX has a purpose,' said James. 'It's not just a camera for the sake of a camera. There are other colour cameras out there too. The difference is that AuroraMAX comes with a dedicated outreach programme.' Volunteers from Astronomy North provide local support and a portal for questions and engagement. And rather than just the black-and-white images of the scientific all-sky cameras, the AuroraMAX photos are in full-colour HD.

The original AuroraMAX camera is a modern version of the THEMIS cameras, but James now uses an SLR camera too because the high-resolution pictures are so popular. They still keep the original camera running because it has more scientific value than the SLR, so both have to fit in the dome. James told me how camera technology has been changing so much, even in the last five to ten years. They used to take eight- to twelve-second exposures but now the sensitivity has improved so much that they can take exposures of only two or three seconds. This means that we can now see brilliant detail in the aurora structures that previously would have been blurred out.

James does all his Astronomy North and aurora work voluntarily, and he clearly loves it. I asked him about his motivation. He puts it down to his passion for science and his desire to contribute.

'If there's a way an amateur can contribute to science I'll try to find it. Up here, by volunteering in this area I can make a contribution.'

James spoke of scientists like Eric Donovan and his team who are making great strides in auroral research. He sees it as a huge opportunity to be able to support and promote the science. 'I'm just a guy who sticks cameras in the snow and makes sure they keep operating all night,' he said modestly. 'It's not my job to analyse the data but it is my job to capture it. As a volunteer, how many times can you say that an amateur astronomer can play a lead role – any kind of role – in a science project of this nature? And all we're doing is taking time-lapse photographs. If observation is the key to making discoveries then I will do it!' He looked at me and grinned. 'I will go out and watch the aurora for the greater good,' he joked. We both laughed.

Yet although the idea of watching the aurora night after night sounds wonderful, I think it's a harder job than James makes out. He endures freezing temperatures and loses sleep, gladly, to bring these images to the public. He is also the one to drive out to the observatory in the middle of the night if there are any issues, such as technical glitches or even clearing snow from the dome if the heater is overwhelmed. It is his responsibility as the president of Astronomy North and he takes it seriously. More than that, he is proud to do it.

It was hot in the observatory. I didn't need the extra warmth of my jumper, but I needed its protection from the bugs. Some mosquitoes had got into the shed as we entered, so every now and then our conversation would be interrupted as one of us slapped at one.

As we talked about James' commitment to AuroraMAX I remembered our discussions on the first night, when we had driven out to the disused gold mine. We had laughed about James' addiction to the northern lights. One of the first things he asked me the night we met was whether I had seen the aurora. I told him about my experience in Kiruna and how, although I had seen it, I wanted to see more, to see it better. I longed to see the whole sky fill up with light above me, rippling and twisting. Yet despite the sky's relative tranquillity in Kiruna, I had still felt grateful to see it.

'I find it fascinating,' I had told James, 'how these coloured lights are able to stir something deep inside, even whilst scientifically I know exactly what is causing them.'

James let out a whooping laugh. 'Well now you're hooked, is what that means!' he cried. 'Welcome aboard!' He told me how he had seen thousands of aurorae and yet still each one grabs his attention. 'What you experienced,' he continued, 'is probably the most common reaction to the very grand event that you were seeing unfurled before your eyes. You were struggling with trying to describe how it makes you feel. Whether you are a scientist or not, as a human being you have observed something incredible, and there's no way you can explain it with a few simple words.'

'It's strange because you don't know how you will feel until you see it,' I said. 'It's nothing like seeing it in pictures or video, even though the modern photographs we see of the aurora are incredibly detailed and vibrant. There is something so different about being there.'

James believes that is why AuroraMAX has captured the attention of scientists as well as amateur astronomers, and indeed anybody who has an interest in nature. It goes back to our ancient connection with the sky. To see it speak with colour and shape and form can only connect us more strongly.

'It's almost as though you're being given a gift,' said James of seeing an incredible auroral display. AuroraMAX exists to share that gift. 'We're saying "hey, everybody! Don't forget how incredible nature is." We have something unique in the North and it's not just a Yellowknife thing. It doesn't matter where you go and see the aurora, just go and see it because it's breathtaking!'

When we left the observatory, James had a quick look around for wolves and bears before locking the shed – just in case we had to rush back in. It was strange to me, something I was not yet in the habit of thinking about. I was more distracted by the mosquitoes.

On the way back in the car, James told me about how he wants to build on the success of AuroraMAX by installing new cameras in other locations. This would give viewers the opportunity to watch how the aurora moves on the large scale, the oval pushing south and then moving back as the substorm cycle washes through. Scientific instruments like the THEMIS cameras do this already, but James would like to broaden it out into outreach with new colour cameras. He hopes it would also help Canadians to relate to the phenomenon better, by watching on cameras more local to them and, when the time is right, going outside. As for the rest of the world, they get to watch live feeds during the daytime – they don't even have to stay up all night to see how the Earth is responding to the bombardment from the heavens.

When we returned to the flat that last evening, we were looking at some photographs that James had taken of incredible aurorae above Yellowknife. He told me how, if he hears there's a geomagnetic storm brewing, he goes out to photograph at a spot that has a good view of the southern horizon. There he sets up three to five cameras in the snow, each pointing in a different direction to give a good chance of capturing the best bits of the aurora. Yet he knows that the glorious shots will be in the southern sky when that substorm, or even fully fledged storm, erupts. The display will go on for hours, so James is out there to film a full-night event.

'I keep the cameras going,' he says, 'that's my job. I have a basic understanding of the aurora science but I have a tremendous knowledge of how to keep cameras going in -30°C and that's come after many years of epic fails.'

'I'm still at the epic fail stage,' I laughed. 'Every time I go to see the aurora I post another terrible picture for my friends.'

'Absolutely. It's kind of like losing the big fish,' he said of the inability to capture the scene effectively. 'You landed the fish and then dropped it in the water.'

Despite over a decade of experience, James has been frustrated too on evenings when he has, say, knocked the camera with his mitt when changing a battery and missed that perfect shot. It's natural. He said he still makes far too many mistakes. His advice on this is simple: 'set up more than one camera, you fool!'

* * *

THE COMPULSION WE feel to capture the aurora, to preserve its majesty for posterity, is not a new reaction. Long before photography, the northern lights feature in art, literature and poetry stretching back through the ages. Scientists and explorers as well as artists put pen to paper to preserve their memories of the wonder. Yet, since the dawn of photography, the focus has been not just to capture a beautiful image as a trophy, but as a tool to aid our understanding. This is still true today. There is still much useful information to be gained by studying photographic images. That is why so many all-sky cameras are scattered around Canada, and indeed the world.

In the early twentieth century the emerging field of photography had a profound effect on our understanding and knowledge of the aurora. But though scientists were beginning to use the images to successfully determine the height of the aurora, their understanding of the atmosphere at those high levels, 80 kilometres (50 miles) and upwards, was limited. Based on his expeditions and terrella experiments between 1897 and 1913, Birkeland had put forward his theory – still highly contested in the early twentieth century – that particles are channelled to Earth on her magnetic field lines and enter the Earth's atmosphere, but still no one knew what actually caused the beautiful light effects that we see. It was only the technique of spectroscopy, merged with an entirely new, audacious kind of physics, that solved the mystery of the colours of the aurora.

Spectroscopy was another technique to emerge at the end of the nineteenth century and that changed the face of auroral research. It was used to analyse light and measure the emitted or absorbed wavelengths in an attempt to determine the substances emitting or absorbing the light. While solar physicists were using the new spectroscope on the Sun, leading to the discovery of helium and the extreme temperature of the solar corona, auroral physicists used it to gain an insight into the chemical composition of the aurora. Scientists wanted to understand how the beautiful coloured lights of the aurora came to be. What was happening in the atmosphere to create these spectacular effects? Answering this question proved problematic because the specific wavelengths of the aurora had never been measured on Earth before. Also, at that time, although it was known that atoms could emit specific wavelengths of light, there was no robust theory of how they were able to do so.

The first spectral measurements of the northern lights were made by the Swedish physicist Angström in 1866–7. Spectroscopy allowed scientists to determine the wavelength of the particular colour of light being measured. The aurora produced a line spectrum – that is, it showed very specific bands of colour, not a continuous blend of hues like white light from the Sun. This meant that the aurora shone with its own light, not because of sunlight reflected or diffracted. However, the auroral colours could not be matched to any known elements. Scientists puzzled over this problem for several decades.

As in solar physics and the case of the mysterious 'green line' seen on the Sun (which turned out to be from a highly ionised state of iron), the identified lines were often initially attributed to new elements and were named 'aurorium' or 'geocoronium'. Yes, it all worked out with helium, which was discovered on Earth in 1895, almost thirty years after Norman Lockyer had identified its spectral line on the Sun, but for the other elusive lines it wasn't that simple. The reason the auroral composition was so difficult to identify is that auroral colours are from *forbidden transitions* – they cannot happen on Earth, only in the extreme conditions found beyond our planet.

Pairing spectroscopy with the advancing field of photography improved the studies by enabling spectra to be recorded for later work. In 1912–3, the Norwegian scientist Lars Vegard, a colleague of Birkeland at the University of Christiania, identified the low-level blue and crimson bands as belonging to nitrogen, but the strongest, green, colour was a mystery until 1923. Vegard worked to reproduce the auroral spectrum in his laboratory. Hypothesising that the green auroral line was emitted by excited nitrogen in an unusual and unknown state, and drawing on his experience of crystallography, in 1924 he began bombarding frozen nitrogen crystals with electrons and discovered some forbidden transitions. The light emissions he created had wavelengths very close to the green-line wavelength of 557.7 nanometres. He suggested that the green light of the aurora was caused by incoming electrons exciting minute nitrogen crystals in the atmosphere, at the same time proposing the theory that the upper atmosphere above an altitude of 100 kilometres (60 miles) was composed of frozen nitrogen at a temperature of almost -240°C. Many disagreed with him.

One objection came from the distinguished Canadian physicist professor John McLennan. He and his graduate student Gordon Shrum repeated Vegard's experiments with frozen nitrogen and satisfied themselves that the nitrogen did not produce the correct wavelength of light as seen in the aurora. Shrum performed further experiments over the following months, studying helium instead as per another, more popular, theory of the composition of the atmosphere. As is often the case, serendipity intervened and led Shrum to accidentally perform the experiment one day without properly purifying the gas. It was contaminated with air and, on this day, the green auroral line showed! Further experiments led to the realisation that the green line came from oxygen. Incidentally, it is now also recognised that the frozen nitrogen used in the earlier experiments must also have been contaminated with oxygen to produce the lines that Vegard originally saw in 1924.

Thus by the end of 1925 the green colour of the aurora had been identified as coming from oxygen, but there was still no credible theory for how or why it happened. The new theory was not long in coming. In the previous decades, physics had made the foundation-shaking move from the classical to the quantum.

Over the first quarter of the nineteenth century the new atomic physics was being pieced together. The Rutherford model of the atom was proposed in 1911 after the now-famous alpha particle and gold foil experiment of 1909. Rutherford and his students did an experiment firing positively charged alpha particles* at thin gold foil of just a few atoms thick. They detected how the alpha particles were deflected by the atoms in the foil. They found that while most passed through unaffected, the paths of some were slightly bent and a few alpha particles even bounced straight back, which was unexpected and surprising. This meant that the atom couldn't be like the 'plum pudding' model proposed by J.J. Thomson after he discovered the electron in 1897 – an atom like a positive pudding studded with tiny negative electron-raisins. Rutherford realised that the alpha particles would bounce straight back only if they encountered a strong positive charge (the alpha particle itself is positively charged, and similar charges, like similar magnetic poles, repel each other). Where there was no positive charge the alpha particle would pass straight by. So most of the mass and charge of the atom must be concentrated in a small central body, encircled by electrons a vast distance (relatively speaking) away – a tiny, positively charged 'nucleus'.

This classical planetary model of the atom was itself quickly supplanted by Bohr's quantum model in 1913. A planetary atomic system couldn't be quite right, since circling like this the electrons would gradually lose energy and spiral into the nucleus. In Bohr's model, electrons occupied fixed orbits, also called states or levels, with particular energies. The electrons are not actually 'orbiting'

---

* Alpha particles are in fact helium nuclei. At the time of Rutherford's experiments no one knew the nucleus of the atom existed. However, they did know that alpha particles were positively charged helium – helium atoms they thought at the time.

the atom in the traditional sense but existing in an undefined cloud around the nucleus. In the quantum world particles can be in several places at once. In the atom, the smaller states closer to the nucleus have lower energy. Electrons can move, or transition, between different energy states only by absorbing or emitting energy. That we see the northern lights at all is because of the intricacies of atomic physics.

When charged particles, predominantly electrons, in the magnetosphere are accelerated by the process of magnetic reconnection, this extra energy allows them to penetrate deeper into the Earth's atmosphere than they would do otherwise. The deeper they travel, the more dense the atmosphere becomes and the more chance there is that they will collide with an atmospheric particle. Without these collisions there would be no aurora.

When a fast-moving electron hits an atmospheric particle, say an oxygen atom, it can knock one of the oxygen atom's electrons into a higher energy level – we say that the oxygen atom is 'excited'. When we discussed plasma previously in Iceland, we heard how electrons could be stripped away from the nucleus of the atom in a process called ionisation. In the case of an excited atom, the electron is not stripped entirely, but it moves into a higher energy state, still confined to the atom but occupying a shell further away from the nucleus.

However, this is but a temporary state for the atom. The excitement cannot last forever and all too soon the atom will relax back. At this climax light is released, liberating the extra energy gained from the incoming electron as a photon. This is the light of the aurora, which is made visible to us as billions of these collisions happen simultaneously. One can think of it a little like throwing a stone up into the air. The stone must come down again, releasing the potential energy it was given in the throw. So the elevated electron must come down and release its extra energy.

The amount of extra energy that the electron had determines the colour of the light that is released. The frequency of the light is proportional to its energy. Higher energy light will have a higher frequency; in other words, the light wave will be wiggling faster and so be towards the blue end of the spectrum. Lower energy light will have a lower frequency – will be oscillating more slowly – and so be nearer the red end of the spectrum.

However, the situation is not sufficiently simple that we can say bigger solar storms will deposit more energy in the magnetosphere and so make more green light. It's actually the reverse. Green is the most common colour seen in the aurora, produced by atomic oxygen. Red and purplish-pink colours are seen much less frequently, generally during the bigger storms. This is due to the changing density in the atmosphere and the vagaries of atomic physics.

Each atom has a series of different positions that an electron can take. These are the different energy levels. In everyday life, when not excited, the atom will be in its ground state, meaning that all electrons are occupying the lowest possible energy level in a shell as close as possible to the nucleus. When an electron is propelled into a higher energy level it is like going up a step – it jumps up to a defined position requiring a fixed amount of energy to get there. There is no halfway. The electron is either up the step or it is not. And this step – this position – is exactly the same in every single oxygen atom there is, which means that, when the electron drops back down, the light released from that position will always be the same colour. In this way the aurora has characteristic colours.

However, electrons in a particular atom can take a variety of positions. By gaining different amounts of energy the electron can move up to different steps – fixed steps depending on the particular atom. Atomic oxygen has a green step and a red step, but it doesn't have a blue step and it never will. The blue, violet and pink colours are released by nitrogen molecules, and the nitrogen atoms emit a turquoise-green that may be obscured by the bright oxygen green. Hydrogen gives out a pink-hued crimson.

The dominant colours of the aurora depend on the timing of these various atomic or molecular transitions and the change in atmospheric density with altitude. In the green-step position the oxygen atom relaxes back and releases green light after about three-quarters of a second. This is actually quite a long time in atomic terms. The nitrogen transitions happen almost instantaneously. Conversely, for an electron in the red-step position it takes almost two minutes before light is released. It's a bit like holding yoga poses. The nitrogen transition is like a shaky handstand that comes down almost immediately; the oxygen green transition is like a slightly more stable headstand that nevertheless can't endure for long; the red line is like a relaxed head-down dog pose that can be held for some time before release.

If in the intervening seconds or minutes the excited oxygen atom collides with another, then it will lose its extra energy during the collision in a withering fizz of heat and the colourful auroral climax will never occur. Thus as the density of the air increases the more other particles there are in the way to collide with the excited particles and quench the auroral light. It is because of this time delay that the auroral transitions are 'forbidden' and difficult to see on Earth. Even in our best vacuum, there are simply too many other particles getting in the way.

The long delay in the red transition is why the red colour is most often seen high up in the atmosphere, above 200 kilometres (120 miles). Up there the air is so rarefied that the oxygen atom can travel on average four or five kilometres (3 or 4 miles) without hitting another particle, so the chances of the minutes passing without a collision are greater. Any electrons with enough energy to reach the green step in atomic oxygen will penetrate further into the atmosphere. Green dominates in the intermediate region because, above about 150 kilometres (90 miles), there are more oxygen atoms to release green than there are nitrogen molecules to release blue. The blues and magentas of molecular nitrogen tend to be seen in a low band below 100 kilometres (60 miles), where oxygen atoms combine into molecules and the higher atmospheric density fizzles out any stray, excited oxygen atoms.

Another reason we see more green aurora than red is that the human eye is just not as sensitive to red light as it is to green, particularly at night when our eyes are sensitive to low-intensity light only in black and white. The red emission may be there, but it might just be too faint to make out by eye. Take a long-exposure photograph with a camera and you may well see a red fringe across the top of the green curtain of the aurora. Indeed, photographs can be very much more vibrant in colour than what we can actually see with the human eye. Often witnesses report seeing a white aurora, particularly the tall pillars of light stretching up to the stars. This is when the aurora is too faint for us to make out the colours. The human eye builds up the picture in black and white.

Back in England, I was shown some photographs by Chris Lee of the UK Space Agency, a keen astronomer and photographer. One of the pictures was an original photograph taken in Finland; the other was the same image with the colour saturations manipulated to make it look as Chris remembered the scene. Together they form a representation of the same auroral display viewed by the camera and by the human eye. The image shows flat snow in the foreground and a bright, wooden cabin to the right backed by a forest of short fir trees. Above, a green aurora reaches up into the heavens, showing a slight twist in the centre of view and illuminating patchy cloud in the distance. In the picture representing the human eye view, the colour is washed out to a muted eerie green. The twist in the auroral arc is less obvious and the stretching upwards less pronounced. The whole sky takes on a paler tinge.

It was a new appreciation of the atom, combined with the emerging field of quantum physics, that enabled the interpretation of the auroral spectrum and brought about our understanding of the colours of the aurora. Alongside this, scientists were able to learn more about the composition of the atmosphere by studying these unusual, colourful transitions specific to certain elements.

Using spectroscopy to measure the light being given off in the aurora, they could identify the elements producing the different colours and thereby find out about the upper atmosphere. Scientists began publishing height profiles of oxygen and nitrogen in the atmosphere, using the subtle light to find out about the composition of this entity that was otherwise invisible.

More than that, spectroscopy became a tool for studying fundamental quantum physics. Scientists realised that many of the spectral colours seen in the aurora were not seen on Earth. By knowing that these transitions existed, and by studying them in the aurora, scientists were able to gain fundamental knowledge about the nature of atoms, which they could not have done in the laboratory alone.

Auroral imaging and spectroscopy are still being used as tools to learn more about the solar system and beyond. Just as scientists initially analysed the colours of Earthly aurorae to determine the structure of our atmosphere, now scientists can analyse the aurora on other planets to learn more about those planets. Moreover, now that we can put satellites in space, we can gain information about cosmic objects by imaging in ultraviolet wavelengths, which we would not be able to see from Earth due to the ultraviolet light being absorbed by the atmosphere. Seeing the aurora on another planet, such as Jupiter or Saturn, tells us two things – that the planet has a magnetic field and that it has an atmosphere. Using spectroscopy to analyse the auroral light emission in more detail can tell us about the composition of that far-away atmosphere.

In the Earthly aurora too, auroral colours and patterns can tell us about conditions in the solar wind and magnetosphere. The solar wind particles may have a broad range of energies, all travelling down the magnetic field lines at slightly different speeds. Particles travelling faster will penetrate deeper into the atmosphere and produce green lights, so seeing predominantly red colour indicates that the incoming solar wind particles are relatively slow. There is usually a mixture of red and green because of the range of particle speeds, and studying the ratio of colours can give an idea of solar wind particle energies. With this extensive information to be gained from imaging, it has been proved time and again that auroral pictures can be much more than just pretty.

* * *

BEING UP IN Yellowknife with James Pugsley and discussing the nightly patterns of the aurora highlighted how little I actually knew about the northern lights. I am a scientist and I have an understanding of the physical processes that cause the aurora, but I've only seen it a few times in my life. I can't really say that I know the aurora. That is why people like James, who live in Yellowknife and who have profound experience, can be so valuable to a project. They bring different perspectives and expertise to the discussion. The success of a project like AuroraMAX is partly due to the mix of people involved: a basis of strong, knowledgeable scientists like Eric Donovan at the University of Calgary, run by the dynamic and dedicated James and with support from the Canadian Space Agency.

James has seen so many auroral displays that he knows what to expect each time and this helps him to set up his cameras in the right places to get the best photographs. The regular flow of the Dungey Cycle creates reproducible patterns in the aurora that are recognisable to the trained eye. It may be a turbulent process producing unique displays but the substorm process shows definite commonalities in form. Scientists like Eric in Calgary have studied countless images to track the evolution of the aurora night after night. 'It's a consistent response of a complicated system,' said Eric. 'In some way substorms always look the same. It's like an explosion but they all look the same.'

Substorms have three main phases – the growth, the expansion and the recovery. With a south-facing solar wind magnetic field, magnetic reconnection on the Sun side of Earth opens up geomagnetic field lines and starts the growth phase. As field lines are pulled over the Earth they build up plasma in the tail of the magnetosphere storing up energy behind the Earth. The auroral oval is the boundary between the open and closed magnetic field lines, so as the growth phase continues and more and more magnetic field lines are opened, the auroral oval moves equatorwards. The growth phase can typically last up to an hour and, during this time, observers on the ground in the north will see the east-west arc of light push slowly southwards.

When the build-up of plasma in the tail becomes too much, magnetic reconnection occurs. The field lines tighten towards Earth and particles are catapulted down into the atmosphere. Beautiful, bright aurorae will result on the night side of the Earth, around midnight. To observers below, the very start of this phase is marked by a sudden brightening of the arc, called the onset, and then the homogeneous arc breaks up into multiple dancing auroral forms. From above, the auroral oval appears to bulge out on the night side. This expansion phase of spectacular aurora generally lasts between thirty and sixty minutes.

Next comes the recovery phase, when the magnetic fields and currents in the magnetosphere relax back into their original state. At this point the aurora becomes patchy and blob-like with pulses of light.

The build-up, the explosive expansion and the recovery together constitute a substorm. The substorm pattern was first noticed by a young Japanese researcher, Syun-Ichi Akasofu, from the University of Alaska, Fairbanks, in 1964. He had been studying all-sky camera images made during the International Geophysical Year of 1957–58, which was scheduled over a solar maximum. This was another year of intensive worldwide scientific collaboration. This time a network of all-sky cameras had been set up across the globe, in both the Arctic and the Antarctic. Including observations contributed by amateur enthusiasts, the images gave the most complete picture of auroral activity thus far. Akasofu noticed that occasional but

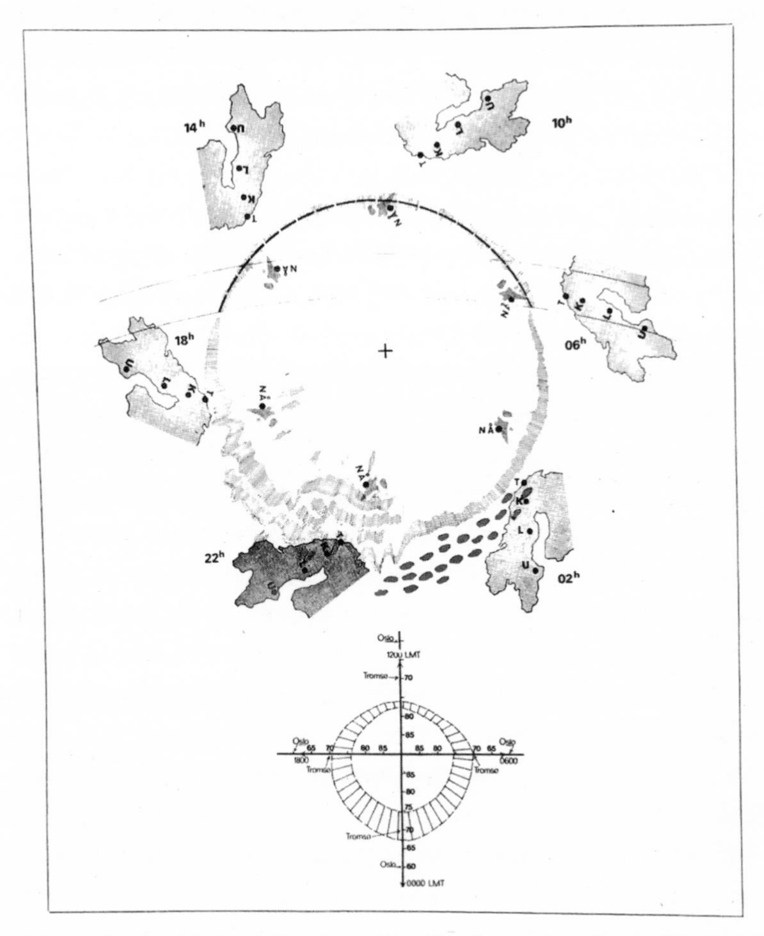

Diagram showing how Scandinavia and Svalbard move in and out of the auroral oval as the Earth rotates, and how the auroral activity varies around the oval relative to the Sun. The Sun is out at the top of the page, so midday is correspondingly at the top of the diagram (closest to the Sun) and midnight at the bottom. Courtesy of Robert H. Eather. Egeland & Stoffregen, Norwegian Academy of Sciences and Letters.

intermittent times of brightness and activity interrupt the otherwise quiet auroral oval. This activity happens furthest away from the Sun, so at midnight. The Earth is constantly turning under this oval, which remains fixed around the magnetic pole, and so a particular point on Earth will experience different auroral conditions as it moves relative to the Sun, moving in and out of the auroral oval as it goes.

I talked with James about his experience of substorms and asked him what he normally sees in Yellowknife.

'"Normal" to us means that we are seeing some nice bright green aurora,' he said. Even when the solar wind and magnetospheric conditions are quite calm and normal, Yellowknife will still have a beautiful, predominantly green, aurora and there will be some sort of break-up. The structures are always similar and they occur directly above. The arc of green light starts in the north, works its way overhead to the south, releases energy in a brightening and breaking-up of the arc, then comes back to the north. The oval expands and then it contracts. If the event is relatively substantial, alongside green tones James may see some of the pinks of nitrogen as the incoming particles penetrate down to 80–90 kilometres (50–60 miles) above the surface of the Earth, where they create some of the more vibrant colours.

Almost every single night they see a similar process occur as the system goes through the same, repeatable phases. It may not be very intense; it may be very gentle. Some nights the auroral arc stays directly overhead, rather than pushing all the way south, and there's no big break-up – just a soft release before the arc transitions back to north. As James said, 'It's as if the wash, rinse, repeat is still happening but it's happening in a very subtle way.'

In Yellowknife, as in other places directly under the auroral oval, all this activity is going on straight overhead. 'That's to me what makes this a unique location,' James told me. 'You're not looking north for the northern lights, you're looking straight up.'

Sometimes, during the onset and break-up, the movement of the aurora can be very fast, with changes happening in the blink of an eye. Cameras cannot capture this kind of movement because they will only get a few frames, but it is spectacular to see. If you are under the aurora when it brightens, the intensity can be such that one can read a book by its light. James often carries a book with him when he is out with visitors. Some nights he can start reading a passage to them while they are staring mesmerised at the sky.

'I'm just trying to prove the point,' said James, 'that this light show can be very intense and it can illuminate the sky. You can suddenly go from pitch black to being able to read a book in just a few seconds. The aurora is very dynamic.'

The camera cannot capture a lot of the fine structure in a very intense, fast event because the aurora is so fast-moving. It blurs out. If anything, this shows the fluid nature of the particles in the sky overhead. Often the aurora might appear like a slow-moving phenomenon, but, speeded up, it has a very fluvial quality. James said that you can observe it as though you would water pouring down a rock or a waterfall into a river. 'It's as though there is a waterfall in the sky and you're directly underneath the waterfall.'

When looking at auroral photography or the all-sky images, scientists and observers like James are not interested so much in the stunning photography, but in what is changing in the structure. This shows where the particles are falling and how they are flowing along the magnetic field lines. It all tells a story of what is going on much further than the eye can see.

Yet there is more to the story. Further twists and complications are provided by the fact that the Sun is a dynamic object, ever-changing, and it can throw much more our way should it choose to. Fast solar wind streams and coronal mass ejections – the huge burstings of magnetic loops that send clouds of hot, fast particles towards Earth – cause geomagnetic storms and vibrant aurora. The substorms may be repeatable and predictable but storms are not. Back in Calgary Eric had told me, 'It used to be thought that lots of substorms make up a storm, but this is not the case. A storm can be two orders of magnitude [or a hundred times] bigger.'

Studying storms takes us into the realm of space weather. During a storm, there is a fundamental change in the plasma environment of near-Earth space which leads to dramatically different magnetic field and current distributions in the magnetosphere. This gives exceptionally enhanced aurora. A substorm is what happens during the normal course of events, whereas a storm is a special occasion from an aurora-watcher's point of view.

I remembered watching a video of a typical storm-time aurora with Eric. It was from on the THEMIS cameras across Canada, so we could watch a mosaic of activity across the map of North America.

'Normally what you would see,' said Eric, 'is a band of aurora and maybe a couple of arcs.'

Not this night. During the three hours of data on this storm, Canada was a sea of seething aurora. Eric pointed out cameras to me. 'See, this here is Athabasca, so there would have been aurora visible in Calgary. And this one here would have been vivid and visible in Toronto.' Yet while this would have been an astonishing event to witness, storms are not much use if you are wanting to learn more about the physics of the aurora or the magnetosphere. Sometimes the quieter events are best. All the traditional processes of substorms happen in storms, too, but they are jumbled together and hard to distinguish. Eric likened it to trying to study water waves during a tropical storm.

As prolonged periods of intense disturbance caused by huge ejections of matter from the Sun or fast solar wind, storms tend to happen more frequently around solar maximum when there are a greater number of sunspots liable to burst. During a storm the auroral oval widens and thickens as numerous magnetic field lines forming the barrier between the Earth and the Sun are broken and particles are accelerated to Earth over a much wider area. This is why sometimes the northern lights can be seen much further south than normal, in places like England or well into the United States of America. During some of the largest solar storms there have been reports of aurorae in Florida and Cuba, or in southern Europe. The southern lights have been reportedly seen as far equatorwards as Queensland, Australia.

The aurora is powered by the solar wind and the magnetosphere is shaped by the solar wind. Since the power source is changeable and the magnetosphere is changeable, the aurora is also ever-changing. Yet it is possible to separate the dynamics in aurora that are caused by changes in the solar wind, as the external driving force, and those caused by the internal dynamics of the magnetosphere itself. The interesting physics issues, where there is still some mystery, are the unpredictable instabilities that happen within the magnetosphere, such as the magnetic reconnection or changing currents that begin a substorm (the substorm problem).

We talk about magnetic reconnection in the tail region being the process that catapults particles into our atmosphere to cause the aurora, yet Akasofu, in an article for the American Geophysical Union's *Eos* magazine in April 2015, writes that the stretched tail region of the magnetosphere may store insufficient magnetic energy to snap back hard enough to create the vivid aurora we see at onset. So is something else happening as well or instead? Again, currents have a role to play, yet even after all these years of study there are still fundamental questions that remain open. We still don't know exactly what it is that causes the flare-up of the aurora in the expansion phase. However, we know the flare-up is not really caused by the solar wind. It is set up by the solar wind as the driving source of energy, but the explosive changes that occur just happen spontaneously, not as a response to a change in the incoming solar wind. 'It's like you pile up some oily rags in the corner and then you get a fire later,' said Eric Donovan of this spontaneity. 'It's not because someone came in and tossed a match, it's because something happened in the system.'

This is where the auroral physics field splits in two – into those who study the direct response to the solar wind and those who look at the fundamental physics of the system that governs the spontaneous evolution.

Back in Yellowknife, James was showing me pictures of the aurora – some more stormy than others.

'So this is an example of a very active night above Yellowknife,' he said as we came across one with many red colours in the southern, bottom part of the AuroraMAX circular view. On this night there was continuous activity throughout the evening. Southern Canadians would have been seeing it further south, if only quite temporarily, maybe for a four-hour period before the oval contracted back to above Yellowknife.

'I see those periods as great advertising,' said James. 'That's essentially the view we have every single night.'

That being said, when the aurora is pushing that far south the displays tend to be more colourful because there is much more energy in the system. 'Generally when people see the aurora in the south they see a lot of colour and when they come north they remark that they see a lot of green,' added James. If you are lucky enough to see the aurora further south it's usually brighter, more colourful and more spectacular. It's just not so reliable.

Storm times also mean that the aurora can occur over much more extended periods rather than just in the hours around midnight, so in the northern winter when skies are dark, people may see the aurora even when going about their daily routine. 'You could be just walking out here, y'know, it might be just after work around six o'clock, and all of a sudden "Oh my God!",' said James. He thinks that's why people end up thinking that aurorae only happen in the winter, because when they are surprised by it they remember it. They don't think about the fact that it continues all year round and that if it is dark enough they could see it around midnight, too.

As part of the AuroraMAX project, James built up a dataset of aurorae above Yellowknife, giving them a score in frequency and intensity to help identify patterns and see what was common and what was unique. The data was gathered initially over three years as we approached the solar activity maximum of 2014. What was clear was that although there was a slight upward trend towards solar max it was only slight. In Yellowknife at any rate, whether it is solar minimum or solar maximum the aurora goes on. James plans to continue the data collection into the next solar cycle.

Through the collaboration with the Canadian Space Agency, the AuroraMAX cameras were also able to film the aurora as the International Space Station was passing above and filming the same aurora.

The town of Yellowknife is visible on the ISS view. Looking at both sets of images, it is possible to identify certain common features. Yet, more than anything, it is a successful outreach initiative, allowing people to appreciate just how tall the aurora is, with the ISS hovering just above the highest aurora. It also highlights the space nature of the aurora, which can be difficult to appreciate from the ground. It took a long time before scientists began to realise that the aurora is not simply an earthbound phenomenon confined to our planet.

Neither is it just a beautiful show. The aurora in our atmosphere is like the image projected onto a screen created by a disturbance further out. There is much more going on that we can't see. Here in Canada we have learnt about what is happening behind the screen – the intricate Tai Chi of the magnetosphere – but what are the ramifications of these fast-charged particles as they come into our domain? Moving charged particles create magnetic fields and electric fields, and the effects of these can be felt on Earth in a variety of disruptive ways. In order to mitigate ill-effects, we try to improve our understanding of the processes involved so that we can make predictions of what will happen and give early warnings to those who may be affected. We are trying to find out how changes in the solar wind and the space environment relate to what we experience on Earth. In effect, we are forecasting space weather . . . and not everyone likes a storm.

CHAPTER SIX

SCOTLAND – THE DARKER SIDE OF THE LIGHTS

I STOOD IN A ridiculously long queue for the car hire desk at Inverness Airport, worrying that if they kept me there too long I would be late for my meeting with Gordon. I had a three-hour drive ahead of me and no idea about the road conditions or the traffic that I might encounter. I had a word with the lady in front of me, left my large orange rucksack on the floor and went to buy myself a cup of tea. The queue stretched out far beyond my bright bag and I wondered how long the people at the end would have to wait.

Finally, my tea finished and keys in hand, I left the terminal building to search the car park for my unfamiliar hire vehicle. Soon I was on the road headed for Caithness, the most northerly county of mainland Britain. I drove out through agricultural land, across long bridges over muddy firths where the tide was out, along roads lined with spiky gorse bushes, some flowering in muted yellows. The road wound through little villages on the way, eventually meeting the sea in a small town called Golspie, with its award-winning beach. I drove along the cliff edge, water to my right, highlands to my left, the low mountains hung with even lower cloud. I looked down on cold dark rocks strewn with seaweed, and the occasional sandy beach peppered with pebbles. As I pushed on further north the roads became quieter until it was as though I was almost alone on miles of open road flanked by gorse bushes and the sea. Eventually I turned off the coast road at Latheron and cut across the heathered heathland of the far north-east of our island towards Thurso. And, yes, I was going to see the northern lights.

People don't generally associate the United Kingdom with the aurora. We think of Scandinavia or Canada or Alaska. But the far north of Scotland, around John o'Groats in Caithness, is at a latitude of 59°N – just a little further south than Oslo or the south coast of Alaska. It may not be quite in the auroral zone, but distant, faint aurorae can most definitely been seen on many clear nights (especially with the help of a camera), and whenever there is a surge in activity the polar lights dance over Scotland as well as southern Scandinavia.

In fact, Scandinavia and Scotland are closer to one another than many realise. Norway is Scotland's nearest neighbour across the North Sea, and the Northern Isles of Orkney and Shetland were easy contact points for early, pioneering Scandinavians with advanced seafaring vessels. Written sources begin to record visits to northern Britain by Viking raiders from the end of the eighth century, initially to the unprotected treasure troves of the monasteries and later to establish control of seaways and transport routes around Scotland, Ireland and northern England.

Scotland must have been an attractive, familiar place to settle for those early Norwegian visitors, the coastline and islands showing a similarity to Norway's west coast. The extent of the Norse colonisation is attested by Norse-derived place names almost as far south as Inverness. The name 'Caithness', for example, comes from 'Katanes', meaning 'Promontory of the Cat People' – the Cat People being pre-Norse inhabitants called Picts. Many of the towns with names ending in *-ster* were once Viking farms, or *bólstaðr* in Old Norse, this element of the name being abbreviated and anglicised to *-bster* or *-ster*.

For a short period in its history, Scotland was Scandinavian. It even features in an Icelandic Saga, the Orkneyinga Saga, which tells the history of conquest in the ninth century and of the Earls of Orkney and their numerous, often brutal, battles. The Earldom persisted there, shrinking out of mainland Scotland over time, until 1468 when the Northern Isles were eventually ceded to the Scottish Crown. However, trade links strengthened and over the following centuries many Scottish communities established themselves in Scandinavia, maintaining their close relationship.

The heathland road brought me into Thurso from the south and I drove through the small town past the large supermarket, across the river towards the church, into part of the high street and round and out to the east towards Scrabster, where my accommodation was located on the edge of town. I could see the peninsula of Dunnet Head, the most northerly point of the mainland, and across the water to the Orkney island of Hoy. It was overcast and the sky was as flat as the firth, both pale shades of grey in the evening light.

Gordon Mackie, Chairman of Caithness Astronomy Group, picked me up from the bed and breakfast soon after I had checked in, having just had time to eat a biscuit from the tea tray and brush my hair. A broad and smiling Scotsman with cropped grey hair and a kindly manner, Gordon is an avid stargazer and auroral photographer. By day he works a little way down the coast at Dounreay, a nuclear reactor development site that goes back to the 1950s and is now under decommissioning. We drove for about ten minutes to the Castlehill Heritage Centre in Castletown, where the Astronomy Group have their meetings. Castlehill was designated a Dark Sky Discovery Centre in 2012, approved as a safe and accessible location with minimal light pollution that is perfect for stargazing. A narrow L-shape of dry-stone walls with a large courtyard garden, Castlehill was seventeenth-century farmstead that has been steadily renovated since 2006 to house an exhibition area and an archaeological research facility. It sits right by the water, a single track running down beside the building to the shore and around past the small, old harbour. It is here, out at the back of the building, where the stargazers gather on clear nights. Tonight it was cloudy, but we had high hopes for the following night.

We walked through the courtyard to the entrance at the north end of the building, in the crook of the L, Gordon introducing me to other members we bumped into on the way in. He had organised an informal and somewhat impromptu meeting on the aurora for my rather hurriedly organised visit. We moved through into the exhibition area where chairs were laid out beside antique objects from the current 'Domestic artefacts over time' exhibition. Lanterns and radios, gramophones and sewing machines lined the room, juxtaposed with the modern projector on the ceiling awaiting an input and projecting blank light onto the back wall. Over tea, coffee and home-made cakes, I shared stories of my northern lights journey so far and the small group shared their experiences of the aurora in Caithness, all to a discussion-inspiring backdrop of glorious photographs and time-lapses taken by group members.

'It's nice and flat here so there is plenty of sky and not too much in the way of light pollution,' said Gordon about Caithness. 'Also, there are plenty of roads around the county so you can get to dark places and enjoy the view.' He said that from most places north of Inverness you don't have to travel far to find a nice dark location. The flat horizon and deep darkness are advantages for seeing the aurora and, as Gordon told us, people are very interested in seeing it, even in Scotland. By far the most popular page on the Caithness Astronomy Group website is the one on viewing the aurora, and Gordon receives numerous queries each year from people asking for advice on seeing a display. There was derisive laughter from the assembly when Gordon mentioned the queries.

'I should say,' Gordon clarified, 'that the ones who contact us tend to be quite knowledgeable. It's the ones who phone up the tourist information and ask what time will they be able to see the aurora on such-and-such a night in three months' time when they will be visiting, where you've got to think they don't quite understand what it's all about.' According to Gordon, the tourist office receives many such requests from some who don't realise the unpredictable nature of the northern lights. Nevertheless, the Astronomy Group welcomes the tourists and many can recall occasions when they have been joined by visitors, some of whom will travel the length of the UK just in the hope of catching a glimpse of the aurora.

If they are lucky, it could be a very worthwhile trip. Only slightly elevated activity is required to see a good auroral display in Caithness and, due to its location just south of the auroral zone, if there is enough activity to see the whole sky filled with light then there will also be colour – in contrast to Yellowknife's sky of green. Seeing the aurora at all is of course much rarer in Caithness than in Yellowknife, not least because of cloud cover, but in Scotland patience is rewarded.

One lady, who had only been living in Caithness for eighteen months, told me of her first experience and how lucky she feels to be able to see the northern lights just by standing out in her garden.

'We went to see *War Horse* one night at the theatre,' she said. 'It was upsetting and I had been crying, so the family made me compose myself before we went outside. But then when we stepped out of the theatre the whole sky was green and I just started crying again. It's just so moving.'

Some nights the aurora fills the whole sky with arcs stretching east–west in discrete bands like a green rainbow. In the big displays they see quite a lot of movement and flickering of the lights. In a few seconds it can cross the sky, like waves one after the other. One of the group commented that, occasionally, there was so much activity they didn't know where to point the camera.

'Sometimes you just need to forget about the camera and stand and watch it,' said Gordon, echoing something that James had said to me back in Yellowknife.

The members of the Caithness Astronomy Group are mostly all keen photographers. They head out night after night, if there is something to be seen, on a personal quest for beautiful pictures.

They joked about their lack of productivity at work after a night out watching the aurora. Many members of the group have had photographs published and are increasingly receiving recognition as more people realise that the aurora is something worth seeing and that they don't necessarily have to travel to the Arctic Circle to do it. In fact, there are some interesting possibilities arising by being further south.

That small decrease in latitude means that astronomical darkness returns a little earlier to Scotland than countries in the auroral zone and so the aurora may be seen in August, the same time of year that noctilucent clouds may appear. Occasionally, observers may be treated to a combined display of bright, wispy noctilucent clouds with the aurora above. The group spoke of a very special display on the night of 4th August 2013, when both phenomena were seen together over many places in the north of Scotland. Gordon remembers the night being unusually warm; he stood and watched in shorts and T-shirt. One member of the group, Maciej Winiarczyk, spent the night photographing from the top of Conachreag Hill back over the heathland south of Thurso. Afterwards he made a time-lapse of the display that went viral on the Internet, showing the interest in this rare event and bringing Caithness into the limelight.

Another of the group had been out during that same display photographing the scene from Dunnet Head. With her was an experienced Japanese-Canadian photographer, Yuichi Takasaka, whose work had appeared in prominent publications such as *National Geographic News*. This photographer was in Scotland visiting relatives and had driven up north when he heard there was a good chance of seeing the aurora that night. When he saw not only the aurora but noctilucent clouds as well, he was dancing around the car park in delight. Even he, with all his experience of the aurora, recognised that northern Scotland could bring something new – the opportunity to see two elusive phenomena together – and, moreover, castles and lighthouses to add foreground interest to photographs. There is even the possibility of catching a meteor trail in your aurora photograph during the Perseid meteor shower in August, as the Earth passes through the debris of Comet Swift-Tuttle. There could be worse places to spend a summer holiday.

These extra, special events keep the group on their toes and always waiting for the next great shot. The aurora can be surprising sometimes, even for those who see it a lot. It doesn't always fit the pattern. Some spoke of a really interesting display where there was no evidence of red visible to the naked eye – only green could be seen.

'With the first photograph I took,' said Gordon, 'I got a real shock because there was lots of red there.' He was surprised because there had been no hint of it to the naked eye. 'It wasn't as if the red was visible as a ghostly white; it just wasn't visible at all. Normally we see white rays and know that they will come out red in a photograph, but that day there was nothing. It was really quite strange.'

As with the enthusiasts I had spoken to in other places, they never seem to tire of watching the aurora. They do, however, become more discerning and less likely to spend time on a low-level event. The older guys reminisced about previous years and past solar cycles when the aurora had been better. This current solar max has been disappointing for aurora watchers.

Solar maximum is when the number of sunspots visible on the disc of the Sun is at its peak. It is when the solar magnetic field is at its most disrupted and brings with it an increased likelihood of solar flares and coronal mass ejections. Though solar activity is said to follow a cycle of eleven years, this is an average value and individual cycles can vary. The years 2008 and 2009 saw the longest and deepest solar minimum in a century, when the sunspot numbers hit record-low levels. Solar physicists worried that the Sun's activity could remain depressed for an extended period, like the Maunder Minimum from 1645–1715, when sunspots dropped to almost zero and auroral displays were almost non-existent.

Fewer sunspots were seen during this entire seventy-year period than are usually seen over one active year, and the reduction in solar output coincided with the Little Ice Age.

The Maunder Minimum was conspicuous after the event by the distinctly low numbers in the sunspot record. Previous vagaries in activity can also be identified by sampling the carbon in tree rings. The ratio of the isotopes of carbon (carbon-12 and carbon-14) in carbon dioxide in the atmosphere varies with solar activity; cosmic rays entering the atmosphere create carbon-14, but high solar activity and a lot of solar wind plasma passing around the Earth restricts the influx of cosmic rays and leads to lower levels of the isotope. It can therefore act as a marker of solar activity independent of sunspot number.

John Eddy, in a paper for *Science* in 1976, re-examined solar activity over the Maunder Minimum period in the light of better auroral evidence and the new technique of carbon-14 in tree rings. He found that the dearth of sunspots did indeed accord with a reduction of solar activity seen in the carbon isotope ratio. Additionally, there was a previous, smaller minimum from the mid-fifteenth to the mid-sixteenth century, and a large intensification of solar activity in the twelfth and thirteenth centuries. This Grand Maximum is reflected in the historical record of observed aurorae, which jumped abruptly at this time, and in reports of naked-eye observations of sunspots from China. It was a time when auroral displays were recorded more systematically by what have been termed the first authentic auroral observers. A famous Norwegian philosophical and political work, called *Kongespeilet*, or *The King's Mirror*, also appeared in the middle of the thirteenth century. Written in the form of advice handed from father to son, it provided – as just one of the topics – information on auroral forms, the best viewing conditions and three alternative possibilities for its origin.

Comparison of the Grand Maximum with conditions today is, unfortunately, difficult. Tree-ring data since the late nineteenth century is unrepresentative because the increased atmospheric carbon dioxide introduced through burning fossil fuels interferes with the natural ratio of the carbon isotopes. The steep reduction in carbon-14 since this time – if it were natural – points to an incredibly high increase in solar activity, which we know through modern studies of the Sun not to be the case. However, assuming that the low levels of carbon-14 back then were natural (industrial pollution was insignificant), this indicates that during the twelfth-century Grand Maximum there may have been higher sunspot activity than we have seen in the entire history of our observations over the last three and a half centuries.

If anything, these changes in solar activity over time serve to remind us that what we think of as 'normal' may be nothing of the sort in the Sun's lifetime, just a label that we have given to the most common state of the Sun in our recorded history. John Eddy sees it as 'one more defeat in our long and losing battle to keep the Sun perfect, or, if not perfect, constant, and if inconstant, regular.' The Sun is none of these things, and thus neither is the aurora.

Fortunately for solar physicists and aurora watchers today, after the low in 2008–9 the sunspot numbers did creep back up, but the current cycle, called cycle 24, is one of the weakest on record. Scientists think that we are past the peak now, yet the crest of the sunspot number is only just over half of the maximum of the last cycle. NASA is calling it a Mini-Max. The likely peak was in April 2014, though this can only be confirmed in hindsight with more data.

Weak cycles, however, can still feature large, sporadic events. One of these in this cycle, a coronal mass ejection that occurred on 23rd July 2012, is believed to have been as significant as the huge event of 1859, when Carrington saw the first solar flare. This is the benchmark for big events. That day in July 2012, plasma was hurled from the Sun at 3000 kilometres (1860 miles) per second, more than four times faster than a more usual ejection. Fortunately the speeding plasma missed the Earth, where it could have wrought havoc, but it did hit a NASA spacecraft called STEREO-A, enabling researchers to analyse it.

At the Castlehill Heritage Centre the Caithness Astronomy Group reminisced about previous solar cycles.

'In 2003 the solar maximum was so good that you didn't have to try too hard to see a good aurora every few weeks or so,' said Gordon. He spoke of photographing displays from the back garden.

Others chipped in.

'This cycle has been weaker, we don't see so much. It's maybe a third of the rate of last time.'

'Back in the 70s we used to get aurora far more. The solar cycle was much stronger.'

'And the displays were more intense then. So there have been fewer and weaker displays. It's a shame.'

'I remember back in 2002 or 3 and in the early 90s that there would sometimes be good aurora every night for a week. We got multiple CMEs* and flares.'

'These guys are spoilt,' Chris Sinclair, one of the younger attendees, whispered to me in a low, Scottish lilt. 'They don't appreciate it. Half the displays where I think "ah, that's quite nice", they just say, "that was pretty rubbish".'

I chuckled. 'Yes, I'm still at the early stages,' I replied. 'I would love to see some of the bigger displays where it fills the whole sky, but I haven't been that lucky yet.'

Lower levels of light pollution also contributed to better auroral displays in the past. One of the group told us about a night when there was an electrical blackout. With all the lights in the region out, he could see the Milky Way and more stars than he had ever seen.

An elderly gentleman spoke up. 'When I was a kid the street lights went off at midnight so everything went pitch black. People in the towns nowadays have no idea what it's like to look out of their window and actually see nothing. But you can see the stars.

---

* Coronal mass ejections are often referred to as CMEs.

That's another reason why we saw more aurora back then, because even the smaller displays were more visible and vibrant because there was no light to kill them. More power cuts is what I say!' The group laughed in agreement.

Castletown now has new LED streetlights that direct all the light downwards and reduce light pollution. They say that from an upstairs bedroom window in town one can now see the stars. The astronomy group are happy but they say that there are some locals who now don't like walking their dogs at night because it is eerie under the whiter light of the LEDs. It is a difficult balance. The streetlights are necessary but they do mask so much. In Norwegian towns, and even in tiny villages like those I visited in the far north, the streetlights are often so bright that they obscure all the stars and certainly the aurora. The long dark winters are spent in an isolated pool of light.

Gordon changed the slide to show another of Maciej's time-lapses set to a backdrop of stirring music. Ripples of colour streaked across the screen.

'That was February 2014,' said Gordon, 'and that's the most colourful display I think I have ever seen.'

'I was away again for that one,' Chris said to me. 'I always miss the good ones.'

'So are you on rigs or on boats?' I asked. Chris had told me earlier that he often worked in the north, off the coast of Norway, as a hydrographic surveyor for the oil and gas industry.

'I'm on ships,' he replied, 'which is good because I don't get stuck in one place. Initially I was travelling the world – Africa, Middle East, Far East. But as you get better you tend to come back home.'

'And what do you do out there?'

'We do sub-sea inspections with ROVs,' he said. (That's remotely operated underwater vehicles that are tethered to the ship and operated from onboard.) 'We look at the sediment,' Chris continued, 'and also do inspections to check for damage. But I'm more senior now so I don't get to do any of the fun stuff; I just do the paperwork. It's a lot more responsibility because mistakes are costly.'

During our conversation about his work, Chris touched on a problem – GPS and the need to get accurate positioning for their inspections. He told me of a night when he was on a boat travelling up towards Svalbard from Hammerfest, the most northerly town in Norway. There was a large auroral display and they couldn't work because none of their instrumentation would function correctly. Chris, of course, enjoyed the interlude.

'It was really visible,' he said of the aurora. 'We could see it building up, the columns of light in a dark blue sky. There was a full moon on the horizon as well so the sky was bright, but we could still see the aurora.'

Now I was thinking about Svalbard and the skiing I would be doing next year.

'It was good to see the reactions of the guys on the boat who hadn't seen the aurora before,' continued Chris. 'We had some workers from the Philippines with us . . .'

* * *

CHRIS'S STORY OF the aurora stopping work on the boat highlights just one of the disruptive effects a geomagnetic storm can have on modern technologies. We call the effects of the Sun's activity on the near-Earth environment space weather, and if we can see the aurora in the UK it is likely we can be affected, too. In recent years awareness of space weather has been increasing and it is now on the UK government's Risk Register due to the possibility for severe damage in the case of a rare, huge solar storm like the Carrington event in 1859.

As Earth weather is the changing environmental conditions in the lower layers of our atmosphere (the troposphere and the stratosphere), so space weather is the changing environmental conditions in near-Earth space. This includes the upper layers of our atmosphere (the ionosphere and the thermosphere) and out into the solar system. Space weather thereby encompasses the solar wind – a flow of charged particles – and activity on the Sun that changes the interplanetary environment, but it is particularly concerned with conditions closer to Earth, within the magnetosphere.

When we talk about space weather we use similar terminology to when talking about terrestrial weather. Rain, snow and other forms of precipitation on Earth are all atmospheric water vapour that condenses and falls; in space we talk of particle precipitation as particles fall or are accelerated into the upper atmosphere (causing the aurora). Large, disruptive events are called storms, and we encounter both radiation storms and geomagnetic storms on Earth. While the Sun's energy drives winds, rain, thunder and lightning, the Sun's activity drives solar winds, high-energy particles, coronal mass ejections and solar flares. Forecasting and prediction of space weather are developing in the same way as for terrestrial weather, learning from experiences and adopting some of the meteorological techniques.

Under the main umbrella of space weather there are three different solar phenomena that can affect the Earth in different ways: solar flares; radiation storms; and geomagnetic storms. The Sun is always releasing optical and near infra-red light. It brightens our days and warms our planet. During solar storms the Sun can also send out extreme ultraviolet X-ray and radio wavelengths of light, plus high-energy particles, which are not always so welcome. In addition, the Earth may be bombarded by fast-moving plasma thrown out from the Sun in a fierce coronal mass ejection. Extreme space weather is usually associated with such an event. The plasma may move at speeds above 800 kilometres (500 miles) per second, pushing a shockwave ahead that compresses the normal solar wind, accelerating it and changing the orientation of its magnetic field.

If the field turns southwards the solar event becomes geo-effective, meaning we will get strong geomagnetic storms on Earth. The southward magnetic field allows reconnection to take place on the day side of Earth, whereupon energy and particles can enter the magnetosphere where they are stored up in the tail region and released explosively as the climax of a substorm, causing beautiful aurorae on the night side.

The Sun sometimes treats us to multiple coronal mass ejections in quick succession. If this happens, the subsequent plasma blobs slipstream the earlier ones, catching up and interacting. Intense geomagnetic storms ensue as the plasma interactions alter the magnetic field configuration over a protracted period. The substorm cycle will repeat every one or two hours while the magnetic field of the incoming plasma remains southward. Thus a geomagnetic storm can contain multiple substorms, and extreme events can last several days and consist of pulses of higher activity within periods of relative calm.

The history of space weather could be said to go back to the Carrington event in 1859, when the British astronomer Richard Carrington observed a solar flare while studying sunspots. Instruments at the Kew Observatory also detected disturbances in the Earth's magnetic field minutes afterwards, and bright auroral displays were witnessed across the world in the following days, some sightings recorded as far south as Cuba and Hawaii. There were reports of unusual behaviour of the still-young telegraph system, which came into use at the end of the 1830s as a means of sending a message in code, such as Morse code, via electrical signals. By 1859 there were telegraph wires stretching across countries and continents, the first undersea cable had been laid between Britain and France almost a decade earlier, and the Atlantic Telegraph Company in London was planning to lay a cable across the ocean. For the first time in history here was a system susceptible to the changes in magnetic field occurring during a large solar event, and the disruption was obvious. Some operators received shocks from spark discharges (like mini lightning bolts), which also set telegraph paper on fire. Even when they disconnected the batteries, currents continued to flow.

Another large event occurred in May 1921. On this occasion, a telephone exchange burned down in Sweden and the New York railways lost some signalling systems. People began to realise that there may be a darker side to the northern lights.

By this time, radio operators already had an inkling that the Sun was having invisible effects on Earth. Radio developed through the early twentieth century, beginning when Guglielmo Marconi succeeded in sending a Morse Code radio signal – the letter S – across the Atlantic from Cornwall to Newfoundland in December 1901. Around the same time, the English mathematician Oliver Heaviside and American electrical engineer Arthur Kennelly independently suggested that radio waves could bounce off the ionosphere, thereby enabling communications around the curvature of the Earth. There were various developments over the first two decades of the century in generating, amplifying and detecting radio waves. By 1910 radio waves were in use as communication between ships and land stations and they began to be used for the emerging aircraft, too. In 1915 the first speech signals were transmitted from Arlington, Virginia, to Paris, where they were picked up by receivers on the top of the Eiffel Tower. Scientific radio experiments had been conducted from the top of the Eiffel Tower for about a decade by that time, proposed and initially funded by the architect Gustave Eiffel as a means to protect his creation from destruction. Had it not proved its worth in this way, the tower would have been dismantled in 1909, having served its twenty-year lifetime as a commemoration of the centenary of the French Revolution. I, for one, am pleased it still stands. I have always enjoyed the change in perspective that height grants and the view of Paris is no exception.

In the 1920s, Edward Appleton conducted experiments which verified the existence of the Kennelly-Heaviside layer of the ionosphere, confirming the existence and the height of the theorised location where the radio waves were bouncing. He also found a higher, more richly ionised, reflecting layer. Appleton's work improved our understanding of the ionosphere and contributed to the development of long-range radio communication. For this he was awarded the Nobel Prize in 1947.

Throughout the 1920s, as radio progress and use was accelerating, people began to notice disruptions in service – hissing sounds or even the signal fading to nothing. Scientific studies determined how radio waves propagated through the atmosphere and also identified how disturbances in the ionosphere could impact radio transmission. Marconi realised the connection between solar activity and the radio disturbances. He even had monitoring antennae mounted on the top of the Eiffel Tower and issued the first forecasts in 1933 under the auspices of the Union Radio Scientifique Internationale. This was the earliest form of space weather prediction. In May 1935 a huge loss of radio communication coincided with the observation of a large solar flare – here was a clear link between a particular solar event and the effect on Earth. The onset of the Second World War and the necessity for reliable radio communications drove the development of forecasting capability.

In early 1940s wartime, the British formed the Inter-Service Ionosphere Bureau and the US formed a similar agency, both aimed at increasing the understanding of radio propagation, collecting ionospheric data and providing timely reports on radio quality conditions. Some predictions were possible, particularly around solar maximum when activity was strong and repeatable, but even real-time alerts were valuable. With a technology in its infancy there are often issues and failures, so radio disruption was often attributed to equipment failure. Being alerted to the potential for radio trouble, operators saved time, energy and money in investigating equipment and making unnecessary repairs.

After the war, countries saw a need for increased collaboration, not just the established scientific collaboration but rather for political partnerships to regulate the burgeoning international air travel, communications and new opportunities for trade and commerce.

Internationally, various councils, committees and agencies formed, reformed and amalgamated over the years, each taking responsibility for a different piece of the global puzzle. Space weather was one of these pieces. The US radio communications work was declassified and their agency for ionospheric monitoring became the Central Radio Propagation Laboratory, located in Boulder, Colorado, and an early incarnation of what would one day become the Space Weather Prediction Center, the world's foremost space weather forecasting centre. Then the fear-driven competitiveness of the Cold War accelerated technological advance in aeronautics and the space age began. Suddenly we were able to explore and study an environment *in situ* where previously we had only been able to observe from afar.

In January 1958, Explorer 1 launched the US into the space age, just slightly behind the Russians whose satellite Sputnik blasted off into space in October 1957. Explorer 1 was the very first US satellite in space, the payload of a Juno 1 rocket that achieved lift-off from Cape Canaveral, Florida, on 31st January. The atmosphere was tense – this was the second attempt to launch a satellite. The first attempt was a Vanguard rocket that had lost thrust and toppled over on the launchpad two seconds after take-off just three months earlier. However, the Explorer 1 mission was a resounding success, not just in being the first US satellite in orbit but also as a science mission.

One of the instruments that Explorer 1 had onboard was an energetic particle detector made by James Van Allen from the University of Iowa. It was simply a Geiger counter attached to a miniature tape recorder. Explorer 1 flew in an orbit that looped out like a hula-hoop round the hips of the Earth, at its nearest point almost grazing the Earth at 340 kilometres (211 miles) distant, at its furthest flying 2515 kilometres (1562 miles) away. It made twelve and a half orbits per day. As the satellite moved further away from the surface of the Earth, the particle detector saw that the counts of energetic particles tended to increase. But within this trend, periodically, the signal would mysteriously drop to nothing.

The detector had not failed; it turned out that it had overloaded. Van Allen and his team realised this, did some laboratory tests to confirm the behaviour of the detector, and deduced that there was a band of highly energetic particles stretched around the Earth. These were regions of high-intensity radiation, as Van Allen described, '1000 times as intense as could be attributed to cosmic rays.' Further missions such as Pioneer 3 and Pioneer 4 in late 1958 and early 1959 enabled the scientists to better map out the belts, finding inner and outer zones. They were most intense over the equator and petered out at higher latitudes. Think of the radiation belts as two concentric squashed doughnuts encircling the equator or, in a two-dimensional side view, as widening crescents to the side of the Earth, not reaching up as far as the poles. These are regions where fast-moving particles are trapped by the Earth's magnetic field, and they became known as the Van Allen radiation belts.

Interestingly, fifty years earlier, Størmer, the young mathematician contemporary of Birkeland at the University of Christiania and pioneer of auroral photography, had predicted this very phenomenon. Størmer was inspired by Birkeland's terrella experiments and theory that charged particles followed magnetic field lines to the Earth's poles. He studied mathematically the motion of the charged particles, building on the work of Irish physicist Joseph Larmor from the previous decade. In the 1890s, Larmor showed that charged particles in a magnetic field would travel in a helix, spiralling around magnetic field lines in a stretched-out circular motion. Størmer wondered what would happen if the magnetic field was changing rather than constant, such as in the bar-magnet dipole arrangement of the Earth.

The magnetic field of the Earth is stronger at the poles than at the equator. Just as the lines of longitude on a globe get closer together at the poles, so too do the lines of the magnetic field. Størmer found that in a stronger magnetic field the charged particle would make tighter spirals. In the case of the Earth's dipole, where the magnetic field gets progressively stronger, the spirals of the charged particles would get progressively tighter, like a spring being compressed, until the spring could be compressed no further and the particle would bounce back in the other direction. The strong magnetic field at the Earth's poles acts as a mirror for charged particles like electrons and protons, trapping them in perpetual bounce motion from pole to pole. These trapped particles form the Van Allen radiation belts.

Pioneer 4 in particular showed an interesting additional result completely by chance. It was much like Pioneer 3 in aims – extending the previous satellite's work – and with an almost identical radiation detector, but the two craft saw very different conditions in the near-Earth environment. The days preceding the launch of Pioneer 4 had been highly geomagnetically active, with five consecutive nights of strong auroral activity, whilst before the launch of Pioneer 3 it had been very quiet. Pioneer 4 found the radiation belts to be swollen, with a much greater quantity of high-energy particles than usual in the outer zone (particularly the 60,000–90,000 kilometres/37,000–56,000 miles region). Van Allen commented in a *Nature* paper on the subject that 'Pioneer 4 observations provide the most persuasive, direct evidence thus far available for the solar origin of (at least) the outer radiation zone.' Activity on the Sun was swelling the radiation belts. In 2012, NASA's recently launched Van Allen probe satellites found evidence that during very high auroral activity a third radiation belt can emerge.

In those early years, the Americans launched several satellites in quick succession. The satellites then were much simpler than modern ones and so significantly quicker to build. Five were launched in the first half of 1958 alone (two didn't make it). Towards the end of 1958, the National Aeronautics and Space Administration (NASA) was created, subsuming the National Advisory Committee for Aeronautics (NACA), which had been established in 1915 'to supervise and direct the scientific study of the problems of flight, with a view of their practical solution'. 1915 was only twelve years

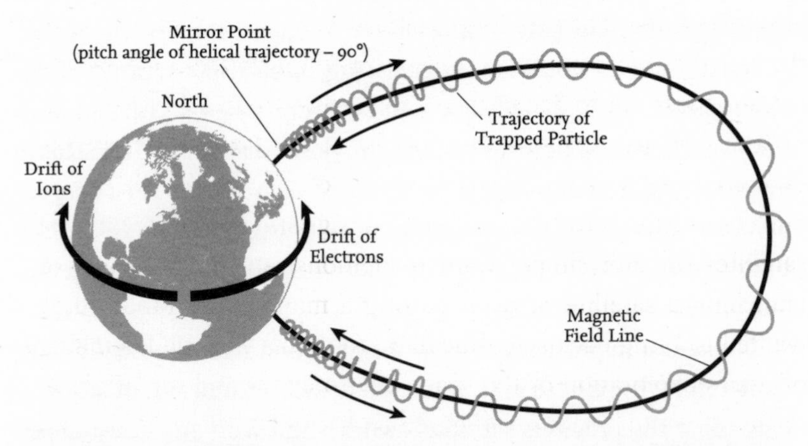

Charged particles trapped in the magnetosphere will bounce back and forth between the poles, following magnetic field lines. Particles in the Van Allen radiation belts are doing just this. 'Pitch angle scattering', as described in Chapter 8, causes the particle to travel more directly down the field line, as if the whole slinky spring has been stretched, so that the particle penetrates more deeply into the Earth's atmosphere.

after the Wright Brothers' first flight but a time when war was raging in Europe and stimulating technical progress abroad.

Early space missions were all categorised scientifically under the heading of 'space science', but as more discoveries were made the field began to diverge into heliophysics, studying the Sun–Earth system, and astrophysics, which was looking out at the rest of the Universe. Explorer 10 discovered the magnetotail in 1961, swooping on an elliptical orbit many times the diameter of Earth away from us. It was then that scientists realised that the magnetosphere took up a complex and dynamic shape. Rather than their imagined 'sphere', a magnetic environment taking up the shape of the Earth, they discovered a stretched-out, windsock-like tail streaming out behind the Earth away from the Sun.

Explorer 12 increased our knowledge further, discovering the magnetopause boundary between Earth space and interplanetary space. As it flew out of the magnetosphere it detected a change from a structured magnetic field to a disordered and weaker field, and in the boundary region it detected a flow of charged particles, likely electrons. The mission also improved our understanding of the Van Allen belts, particularly by determining that the high-energy radiation was not so dangerous as to preclude manned spaceflight.

The frenzied, exploratory progress of the Space Race greatly enhanced our understanding of the space around the Earth and our connection with the Sun. The designers and developers of satellites did not simply want to demonstrate the principle of launching a satellite or even putting a man on the Moon; they wanted to bring back scientific data, too. This was the beginning of man's exploration of the space environment and the origin of an awareness of space weather – the fact that there are things out there in the near-Earth environment that can affect the planet and us on it.

Mariner II, launched in August 1962, was the world's first interplanetary spacecraft. It made a flyby of Venus and sent data back on the planet's temperature, atmosphere and (undetectable) magnetic field. It also sent back important news about the space between planets. It was not 'empty space' at all; rather, it was filled with plasma travelling with a bulk motion of hundreds of kilometres per second. Mariner II verified the existence of the solar wind, an idea that had first been put forward by Birkeland in his major work *The Norwegian Aurora Polaris Expedition 1902–1903*. Only decades later was it taken up by scientists like Chapman and Ferraro, who proposed the concept of the magnetosphere, and Gene Parker, who established a theory of the solar wind and how it spirals outwards from the Sun.

In the 1970s the National Oceanic and Atmospheric Administration (NOAA) was created and the space environment came under NOAA's research remit. But the term 'space weather' didn't come into usage until the 1990s, when it was realised that the changing conditions in the space environment needed continual and coordinated monitoring to protect the commercial and military systems that can be affected.

As time has passed and the world has become more and more

technologically advanced, other countries have come to realise the importance of space weather and the drastic implications there can be for human safety, our livelihoods and our economies. For decades there have been scientific collaborations between nations and forecast data has been available worldwide to those who need it. But the importance of space weather is only gradually filtering down into government.

* * *

'IT ALL GOES back to 2010,' said Mike Hapgood, referring to the UK government first beginning to take a real interest in space weather. Mike is Head of Space Environment at the Rutherford Appleton Laboratory, and he chairs the expert group that provides space weather advice to government. 'What set it off was the volcanic ash episode.'

As we found out in Iceland, the volcanic eruption of Eyjafjallajökull caused havoc globally as flights were grounded. The International Air Transport Association estimated revenue losses of US$1.7 billion over the seven-day period in April, with over 1.2 million passengers affected each day. The UK, relatively close by and right in the path of the ash cloud, was particularly hard-hit.

'It was a bit of a wake-up call for Civil Contingencies,' continued Mike. 'Suddenly this thing came out of nowhere and caused them great problems. Then when they started talking to the scientific community studying volcanoes, the feedback they got was "we've been trying to tell you this for years!".' Mike explained how within the volcanology community there was an awareness of the risk of a potential volcanic eruption and the dire effects it could have, but this awareness had not made it down to the government risk managers. It prompted those working for the government in Civil Contingencies – the department of the British Cabinet Office responsible for emergency planning – to ask what else was out there that they needed to worry about. So they quickly started surveying the various risks, and one that particularly stood out was space weather.

In May 2010, less than a month after the eruption, there was a general election and a new coalition government came to power. Very early on, the Chief Scientific Advisor, then John Beddington, had a meeting with the new prime minister, David Cameron, to discuss the main scientific matters of importance at the time. Space weather was amongst these, and that it was being mentioned to the prime minister meant that the space weather threat had already gained significant momentum within government. By the following year it was featuring in the National Risk Register, the government's assessment of the risks of potential civil emergencies.

When considering the risk of a civil emergency, which the government defines as a situation that threatens serious damage to human welfare, the environment or the security of the UK, it evaluates two things: the likelihood of the situation occurring; and the potential destructive impacts if it does. In 2015, the likelihood of a severe space weather event in the following five years was rated at between a 1 in 2 and a 1 in 20 chance, which is the same as the likelihood of heat waves or heavy snow. This may not necessarily be a worst-case scenario space weather event, but it could be one that could cause significant problems on the scale of the Eyjafjallajökull eruption.

The potential financial impact of a severe geomagnetic storm is difficult to estimate, but the US National Research Council took up the challenge. They established that the wider societal and economic costs of such an event could be between one and two trillion US dollars.

The Carrington event of 1859 is taken as a worst-case scenario when assessing impacts and is considered to be a 1-in-100-year event. It is the largest ever recorded space weather event and is thought to be up to ten times larger than anything seen in the past fifty years. Scientists have estimated that the probability of a Carrington event happening is about 12 per cent per decade. But it is a truly random occurrence – just because we haven't had such an event in over a hundred and fifty years doesn't mean one is overdue and therefore more likely. There is still only a 12 per cent chance of it happening over the next decade.

However, the 1921 event that caused trouble in Sweden was, according to Mike, approaching the size of the Carrington event. 'It adds weight to the assessment of major solar storms as 1-in-100-year risks,' he said.

Another thing to note about both the Carrington event and that in 1921 is that they occurred in otherwise quiet periods of the solar cycle rather than at solar maximum, showing us that while the Sun is more active in general at solar max – and thus space weather effects are elevated – an extreme event can happen at any time. So we need to be prepared. As the years go by, the risk intensifies. The more we rely on advanced technologies, the greater our vulnerability.

As a result of those early meetings in 2010, an independent expert group, chaired by Mike, was assembled to provide scientific advice to government on the space environment should the necessity arise. This is the government's go-to group of people if a major event occurs, with the remit to provide the current best information available to guide the government response. In an emergency there is no time to go away and do a scientific study; we need experts to give the most up-to-date information, even if there remain uncertainties and questions. 'The people can change,' Mike told me. 'The whole thing needs to be dynamic so we can respond to the emergency by bringing in the right expertise.'

In 2012 there was a UK House of Commons Defence Committee assessment of the developing threat from electromagnetic pulses, which included the risks posed by space weather. One of the recommendations from this was that the government should ensure sufficient funding and resources for monitoring space weather. They concluded, 'space weather is a global threat and may affect many regions and countries simultaneously. This means that there is scope for mutual assistance, but also that there is no safe place from which it can be assumed that help will come. It is time that the Government began to approach this matter with the seriousness it deserves.'

In light of this, funds were made available for a new division of the Met Office, the UK's national weather service, which was already working in collaboration with the Americans on space weather prediction and forecasts. The Met Office Space Weather Operations Centre, located in Exeter, southern England, was officially opened in October 2014 to provide a continuous, manned service to protect the UK from serious space weather threats. In parallel there has been discussion with some of the key bodies that are affected by space weather, such as the National Grid, satellite operators and airlines.

When a solar event occurs, radio and communications are affected first. Solar flare X-rays travel at the speed of light, arriving at Earth in only eight minutes. When we see the flare on the Sun the X-rays are already with us, disrupting the ionosphere. Next, the radiation storms occur as streams of very high-energy charged particles also disrupt the ionosphere, but in addition can cause damage to satellites and pose a radiation risk to humans lacking the full protection of the Earth's atmosphere, such as astronauts or airline passengers. Depending on their speed, the particles take between fifteen minutes and twenty-four hours to reach us. Finally, if the coronal mass ejection plasma hits us, we get the geomagnetic storm. The plasma can take between one and four days to get to Earth and is only geo-effective if the embedded magnetic field points southwards. The resulting auroral storm accelerates particles into the ionosphere, which can cause problems with communications and navigation, while the changing magnetic fields can disturb electrical power distribution networks.

A high-speed coronal mass ejection with a southwards magnetic field is the most geo-effective combination there is – great if you are an aurora spotter, less good if you run the National Grid.

* * *

THE MORNING FOLLOWING my discussion evening with the Caithness Astronomy Group, I went out with Chris to see some old archaeological sites. Caithness is rich in prehistoric monuments, some so much a part of the landscape that one has to look carefully to find them. There is evidence of human settlement in Scotland from the third millennium BCE, though hunter-gatherers are thought to have roamed the region as far back as 6000 BCE. Remains have been found of Middle Stone Age hunters and fishermen on the west coast near Oban and stretching south. By the early second millennium, settlements stretched up as far north as Shetland. These were farmers who kept cattle and sheep, and who were cultivating cereals and working metals.

In the first millennium BCE the climate became cooler and wetter. Hill forts appeared around Scotland, and brochs in the far north. The brochs were round towers that could stand up to 15 metres (49 feet) in height and were built by early ancestors of the Pictish people (whose name is said to come from the Romans – the Latin *Picti* meaning painted men and derived from their custom of tattooing their bodies).

'The Romans were terrified of them,' Chris revealed. 'They painted themselves blue and fought naked.'

The first century CE saw the Romans push into Scotland, but subduing the hostile natives was beyond the Empire's strategy and means and so Hadrian's Wall was built between 122 and 128 CE as the permanent northern boundary of Roman Britain. From 400 CE, when the Romans began their withdrawal from Britain, there were four peoples north of this frontier: the Britons and the Angles in the south; the Scots, of Irish Gaelic origin, in the west; and the Picts in the far north. This period, only sparsely recorded, was brought to an end by the Viking invasions from the eighth century. The Norse settled in the far north, Caithness to the Shetlands, and the Picts were pushed southwest where they merged with the Scots and unified in the late ninth century to form the Kingdom of Alba, which stretched from the west of Caithness down through much of central Scotland. Impenetrable to the Scandinavians, Alba became isolated as the Western Isles, Ireland and England were settled by Norsemen. In the eleventh century, Alba was united with southern provinces to form the Kingdom of the Scots and eventually pushed the Viking influence out to the Orkneys.

The sites Chris took me to see were stone circles and standing stones thought to date back to the third and second millennia BCE. Some were on the tourist trails and marked with information plaques, while others were what Chris called 'hidden gems'.

One of these gems is one of the largest stone circles in the UK, almost 60 metres (200 feet) in diameter, but it was down in a dip where knee-high, coarse, dry reeds pushed up in rippling clumps through the tufted heathland grass. Gnarled trees and gorse bushes had grown up over the years and a barbed-wire fence passed through the middle of the plot so the stone circle was hard to pick out. It was unusual for a site to be in a valley and not up at a viewpoint.

We walked down the hill to take a closer look, tramping through the grasses and climbing over fences. Chris was tall and helped me avoid the barbs. The sun was out but the wind was cold and I zipped my jacket up higher. Sheep baa-ed in the distance. In the bottom of the valley ran a small stream, the Burn of Latheronwheel. Its meandering path had eroded close up to the circle, within a few metres of one of the large stones. Being a wide circle, the individual stones were far from each other and could easily pass for a single curiosity rather than a part of a larger whole. But they were unmistakable on closer inspection. The large, rough-hewn, flattish stones stood almost as tall as me, erect, regal, poised. Their sharp edges were weathered smooth and their faces decorated with white lichens and green mosses in degrees of shades and textures. Looking around, I could pick out other standing stones to form a pattern.

'Do you know much about trees?' Chris asked suddenly.

'Not really. Why?'

'I've always wondered about that lone tree up there – the one that is green. It's a different tree to the rest, obviously, but I don't know what it is.'

One sole evergreen stood on the hill, conspicuous between the deciduous trees flanking it.

'It's like a cypress or a cedar or something.' I said uncertainly. 'My sister would know.'

'There is a broch up the strath* from Berriedale that is built right by a yew tree,' Chris told me. 'Yews were sacred to people back then. Apparently they used to use the sap from the tree to tip their arrows, because it's poisonous, and they would use the wood, too.' Chris was curious about whether the lone tree here also had an ancient purpose.

We went on to visit Dunbeath Broch and the Hill O'Many Stanes, which – just as it sounds – is a hill with lots of stones. But it is more than that. All the stones are carefully arranged into a fan shape looking due east out across the sparkling sea. This may not even be the full pattern – there are almost two hundred stones there today, but some speculate that there may once have been up to six hundred.

The mystery today is what was the purpose of these sites built with such great labour and care? There is little evidence to say. Did they have a ritual, ceremonial or territorial purpose? Were they a graveyard (if so, where are the human remains?), an observatory or an astronomical calculator? Archaeologists or astronomers pick out alignments, such as the north–south rows of stones or the opening of a horseshoe aligned with the midwinter rising Sun, and an astronomical explanation is suspected, but it remains a mystery. What is fascinating to me is that these primitive people, thousands of years ago, looked up at the very same sky that we do, looking for meaning or pattern. The stars spun and the aurora danced for them just like they do for us today, though arguably the ancient people had more of a connection to the sky than we do, if less of

* A strath is a wide, shallow river valley.

an understanding. It's as if the sky is a bond that links us to our distant ancestors.

Up in Caithness they certainly have a lot of sky. Even without the aurora it can be an inspiration. That afternoon I drove to the far northeast corner of mainland Britain, through the little village of John o'Groats and out the other side to a croft, or smallholding, right on the far edge of our island. An artist couple, David and Kitty Watt, welcomed me into their home and studio. David was tall with bushy hair, large in stature but softly spoken with Scottish vernacular. Kitty was smaller and dark-haired with a clear English accent. Although I knew that they didn't paint the northern lights, I had seen that David makes beautiful use of colour in his paintings and Kitty of light and tone in her etchings.

After greeting me they showed me into a large, barn-type building adjoining the house. It was full of David's pictures. Huge canvasses of swirling colour hung on the walls with others stacked below on the floor. I asked David what inspires him.

'It can just be moments in time,' he said, 'the things you see, a sudden light.' They have an impressive space – a wide, wild view over open heathland, out to sea and across to the Orkneys. 'Living here can get quite dark,' he said, 'especially in the winter. But then you get the most amazing cloud effects and light. You just try to retain it in your head as much as you can and reproduce it.'

He told me how he sometimes uses very rough sketches to capture moments, but how the finished work often looks nothing like the original sketch. Fortunately his studio looks out to sea, so he always has something to paint.

'A lot of what I do is the views from here, or even just wee bits of it,' he explained.

David has painted since he was a child, but it has been in bursts. He and Kitty moved up to Caithness from Aberdeenshire almost twenty years ago because they wanted some land. They liked the freedom of Caithness, the way the landscape is less populated. Kitty is originally from Sussex, the very south of England, but has been in Scotland for over thirty years. There's something about the space up here, she says.

For the first nine years after they moved north David was too busy working on the croft to paint, and he didn't have a studio either. Then he hurt his back and bought a shed, which became the studio, and he started painting again.

'I've been serious about it since then and can't see myself ever stopping again,' he told me. All the pictures around the barn were fairly recent when I visited. The walls were awash with oranges and blues, white foams and shafts of light, wild seas, swirling clouds, seabirds and sunsets across the seasons. It was a beautiful cacophony of colour.

I asked David about this colour and what determines how he will use it. David just laughed a breathy ha-ha-ha chuckle.

'I have no idea!' he replied. 'It just happens.' I was wondering how someone could take something simple like a walk out on the cliffs and turn it into something beautiful and spirited like this.

'Some people are really good with colour,' said Kitty, 'and some people aren't. David is. I think it's an instinctive thing.' To David it seems obvious, but many people struggle with it.

'I think it's developed over the years,' said David. 'I used to just paint in watercolours because I didn't have the space to paint in oils, but now I use oils. The colours are stronger and I'm more confident as I get older, so I'm really hoping that by the time I'm a hundred I'll be quite good.' He laughed that strong laugh again.

Kitty admits to not having the same instinct with colour, yet she has her own distinct style. Kitty is a printmaker who makes delicate etchings with a different kind of beauty, one that is soft and tonal rather than bold and contrasting. It is completely different to painting. She scratches a design into wax on a metal plate, protecting the parts she wants whitest and using acid to etch in multiple stages for different tones. She has to work backwards and imagine how the finished piece will look when it is complete. It is all about light rather than colour, and much of the skill is in the printing. Kitty's work is delicate, restrained and methodical, requiring a very different skill set to David's.

I looked around the room at the variation in the canvasses across similar scenes, at the changes in seasons and weather. Kitty and David told me about how the winters there could be spectacular, with stormy seas, cliffs soaked with spray and huge, billowing clouds like explosions. They spoke of storms coming across the sea and blowing around them on their corner of land, almost as though they are watching it all from the calm.

'Aye,' said David, 'you get the strong tides and that pulls the weather along either one way or the other. So we get reasonable weather and we have this amazing scene going past over Orkney.' It sounded wild and fantastic.

'And the stars here are just amazing, astonishing,' said Kitty. 'Sometimes the air is just so clear.' That prompted me to ask about the aurora and why they don't use it in their art when they have such a wonderful view.

'I think it's because it's just so spectacular and so massive that to try to restrain it into a frame is very difficult without making it look a bit twee,' replied David, who wants more than just pretty northern lights.

David and Kitty acknowledged that they had seen some wonderful displays. They mentioned one in particular which they recalled as a 'pyramid effect' above them, the light shooting up in pale green with a touch of red. They were describing what is classified as a corona, so the auroral arc must have been directly above them.

'And it's the way it moves,' said David, 'like spotlights. It disappears and moves. I think you're overawed by it and just keep looking at it.'

'There's nothing you can do but admire it really,' added Kitty.

'Aye, you can't beat it,' David concurred. 'The ones I've seen in art can never capture it, that feeling . . .'

'Like when there's a thunderstorm and you can almost feel it,' finished Kitty.

I agreed that a live, all-around experience is completely different to seeing even an incredible picture. Yet, of course, the aurora does frequently appear in art. The northern lights have been painted long into the past. Before photography it was only through art that the phenomenon could be captured at all and preserved for posterity, particularly further south where auroral displays were more of a rarity. Historical works in the sixteenth century show the aurora in the superstitious light of the time, depicting it as flames or knives in the sky. Later, scientists often penned their own pictures for their records and works, such as de Mairan's early textbook on the aurora in the 1700s. Nansen produced a woodcut from a crayon sketch he made in November 1893 when he was on the *Fram*, and a painting of a bright red aurora overhead was found amongst Størmer's records in the twentieth century, even though he spent his career photographing the lights.

Art also allows the injection of something unquantifiable into the scene and brings the possibility of a new interpretation. While we may say that a photograph is not a true representation of the aurora – in that it doesn't represent what a human eye sees, even if it does show up what is actually there – art is not necessarily a true representation either. Indeed, that is sometimes the point. Artistic licence may be used to convey some deeper meaning. A good example is the Romantic painting *Aurora Borealis* by Frederic Edwin Church, unveiled in 1865. Jeffrey Love, a geophysicist working on geomagnetism and the aurora at the United States Geological Survey, interprets the work as 'a restrained tribute to the end of the American Civil War.' The colour palate of the aurora in Church's painting emphasises blue, red and yellow rather than the more usual auroral green and red, and the perspective seems not quite right. Jeff believes the depiction of the aurora is used to subtly project subdued nationalism and uncertainty as 'a flag unfurled across a cold and barren landscape.'

I think that the reason David feels as though art can never authentically capture the aurora is because of its movement and the problem of translation to a static page. I said to him that he had managed to convey a sense of movement in water and cloud, so why not the aurora? All of his pictures are very dynamic – they have waves and storms and clouds.

'Aye, so maybe it will come,' he nodded.

'I've been thinking about it, actually, that movement,' said Kitty, 'and wondering how could I achieve it.'

I smiled. 'I don't mean to say that you should necessarily paint the aurora,' I said. 'I'm just curious.'

'Well, it is interesting,' David conceded. 'I do try to get as much movement as I can into the sky.' He sighed. 'I suspect part of the trouble is that it happens late at night, and you're not at your best late at night. I don't want to be thinking about how best to paint it then.'

It is true that such a dynamic event is hard to constrain on canvas, and many of the pictures one sees of the aurora, particularly the early ones, do seem quite static. Yet judging by some of David's other paintings, I think he could capture the sense of movement of the aurora well, should he ever decide to try.

We went through to the kitchen to have a cup of tea. A glossy collie dog lay curled in the doorway, happily raising its head to be stroked as I passed. The Watts' house was in some ways frozen in time. The kitchen cabinets were well-used 1960s style – painted wood with no frills, a bit like those I had seen in village halls in the 80s as a child. There were a couple of stoves that, I learned, burned peat, and a boiler tank stood bare and unhidden in its mustard-yellow insulation. Between the units at the top of the room, pushed up against the window, was a long, dark-wood table with four unmatched wooden chairs. Sun slanted through the glass, brightening the edge of the rough table and bringing out some paler, golden hues. Kitty ushered me to the sunlit seat. Outside a small windmill spun briskly.

As Kitty boiled the kettle on the stove opposite me we talked of the farm. They have a small herd of cattle and a large garden where they grow their own vegetables. There is a stream outside for water and they cut peat for the stoves from the ground a short walk from the house. We would go up there after tea, tramping though bog and purple heather, to what they call the 'peat bank', a small, disjointed scar in a section of hillside. At one side of the scar is original, as yet uncut, land; the other, lower, side is where the peat has been removed, almost like scraping fat from under skin. The scar itself is a metre-long drop down to a narrow strip of exposed, dark peat, before the hillside continues in grasses and heather at this lower level. Piles of fibrous bricks lay piled like cairns at the end of the strip nearest the house.

Twice a year the Watts remove a foot-wide strip of turf from the higher edge and cut down into the peat, which comes out in big slabs like butter. They slice it into bricks and lay it out to dry into solid cakes of fuel which they burn like wood. Once they have taken out the peat underneath they replace the turf. The grass and heather regrows as if healing the cut in the landscape. Looking over the peat bank, twenty years of fuel consumption for heat is visible in the lower level of the hillside where the peat has been removed.

'We're not on mains electricity here so that's our power just there,' said Kitty, gesturing out of the window to the small windmill that was spinning in the breeze. She set down a large pottery teapot and some rough mugs. 'Do you take milk and sugar?'

I was surprised they were off-grid, being only a kilometre or so away from the village.

'At first we thought it was a bit daunting, but you just get used to it,' said David. 'Yes, our lights do go off now and again but . . . ah . . . it's fine!' he chuckled as Kitty poured the tea. 'It's just a bit weird for visitors if they have to sit in the dark and wait for the generator to go on.'

Kitty passed around the mugs of tea and a container of home-made, fruity rock buns. The Watts have batteries to store some of their wind power, and a back-up generator for when they need more. 'Like when we need to hoover,' said Kitty.

Being off-grid, the Watts are much more aware of their power consumption and of the wind – they know what they need to switch off when the wind drops. There is only a certain amount of energy available that can be used at any one time.

'We can just about hoover if it's really windy,' said Kitty, 'but it's a good excuse not to do it!' she laughed. If they had a bigger windmill it would cope better, but theirs is small and was installed almost twenty years ago, back when wind power was not well established and people thought they were crazy. Now large wind turbines feature prominently in the Caithness landscape as wind farms are erected, though there is much resistance to the landscape being blighted – as the action groups put it – for what they consider little electrical gain. They do have a point, and it is clear from the Watts' experience that a balance of sources will always be required to satisfy rising energy demand and a constant need for electricity.

I left the Watts and drove to the end of the road to Duncansby Head, where I took a walk along the cliffs to see the Stacks – steep columns of eroded rock – that feature so frequently in David's art. I walked past the Geo of Sclaites, a deep, narrow inlet flanked by vertical cliffs where gull-like fulmars nested like lovers, breasts and beaks touching. The cliff path dropped down into a dip before climbing again to almost 80 metres (262 feet) high beyond the stacks. The birds swirled above a narrow beach of boulders and rock pools of seaweed as I climbed. The stacks themselves are around 50 metres (164 feet) high, steeply pyramidal and scored in sedimentary layers of sandstone. The sea was calm, and though the breeze whipped my hair against my face I tried to imagine the scene in wilder weather.

After a brief pause to take it all in, I ran back up the path to the car so I could get over to Thurso in time to meet Gordon, Chris and some other members of the Astronomy Group for dinner. I hoped that the sky would stay clear.

CHAPTER SEVEN

SCOTLAND – FORECASTING SPACE WEATHER

After my time aurora-hunting with the Caithness Astronomy Group, I drove the three hours back down to Inverness from Thurso. But before I left I made a detour to Lossiemouth, on the north-east coast of Scotland between Inverness and Aberdeen. This small town of wide skies and sandy beaches is home to one of the Royal Air Force's main operating bases and, wondering about the space weather effects on aviation, I went to meet ex-RAF pilot and now part-time pilot trainer, Al Read. Rather than meeting at the base, we met at the nearby industrial estate where Al runs a micro-brewery, Windswept Brewing Co., with his business partner Nigel Tiddy. Inspired by their northern skies, Windswept released a special limited-edition beer called Aurora – a northern, light pale ale.

As pilots themselves, there may have been a second inspiration for the name of the beer. At Windswept they often call their beers after aircraft. Tornado and Typhoon – two of the jets that are based at Lossiemouth – are two of their beers. And there is a rumour of a top-secret US military spy plane called Aurora, which is an alleged hypersonic aeroplane. It may not have been the original inspiration, but it's an interesting crossover between aircraft and the local themes that Al and Nigel generally use as ideas. 'We have one beer called The Wolf,' Al told me. 'This is named after a local badass, the Wolf of Badenoch. In 1390 he burnt down Elgin Cathedral, amongst other things. He was quite a character.'

With cups of tea in hand we sat down in the small office, a portacabin outside the main brewing shed. Until recently four of them had been crammed into a box-room office in the main building, tucked away beside the brewing room. Now they had more space, a neat carpet and a small shop counter at the entrance to sell to individual customers. There was a large European map on the wall.

Windswept Brewing Co. started at the end of 2012 and has been expanding quickly – in about two years their brewing capacity increased fivefold. Al puts this down to the steady demand for craft beer. 'Over the last few years it has been one of the few growth sectors in the beer industry,' said Al. People have an interest in where their produce comes from now; locally sourced beer, as opposed to the big, mass-produced brands, is increasingly popular. Having said that, Windswept don't just sell locally. It's true that the majority of their sales are currently in the Aberdeen to Inverness area, but they also have some distributors selling their beer into the rest of the UK and they export bottles to five European countries, soon to be six. They may soon be crossing the pond into Canada, where I expect their Aurora beer could prove a hit.

As for many small, growing companies, the To Do list at Windswept just never ends, but the experience and skills that Al and Nigel learned in the RAF prove very useful in this kind of environment. There is a need for prioritisation, organisation and an ability to separate out what is important and what is not. 'There is an air combat term that goes: "fight the closest, hottest threat",' Al told me. 'That's the order in which you hit things out of the way. We're still doing that here.'

Al left the RAF six years ago and went to work for the defence company Thales, just as Nigel had done two years previously. Here they used their own experience of tricky flight situations to create training programmes for pilots in the simulator. The trainees sit in a replica cockpit in a large dome where a computer-generated world is projected around them. They are presented with problems that require them to exercise judgement and build thought processes so that – hopefully – they never see something for the first time in combat. The training is like a big game, but with serious undertones. Al and Nigel are preparing pilots for emergency cases, such as when things are going wrong in their plane.

I asked Al if he had seen the aurora much whilst flying. Pilots must get a good view from the cockpit.

'Yes, I used to see them when I was flying,' he told me. 'I don't remember them ever being spectacular when flying but it's the kind of thing I would often see in the distance, further away to the north. I remember doing a period of night training and pretty much seeing them every evening.'

'Was that up here in Scotland?' I asked.

'Up above the north of Scotland and looking north,' he replied. 'I'd be sitting at fifteen to twenty thousand feet, and with the clearer air up there your visibility is a long way. Then when we were wearing our night-vision goggles . . .'

What a good idea! I thought. Night-vision goggles amplify light and allow the pilots to see in very dark environments.

'It amplifies starlight and any light that's there,' Al continued, 'so quite often when the aurora was too weak to see with the naked eye we would pick it up with the night-vision goggles. Actually I would highly recommend night-vision goggles for aurora spotting. You need to get a set!' he laughed. 'Now they're only monochromatic, they won't give you the different colours, but they'll certainly highlight to you when something's there.'

He carried on to tell me how the night-vision goggles were also great for spotting shooting stars. When Al was working down in the Gulf he used to do quite a lot of night work. He told me how he would fly down the Persian Gulf at four in the morning on a transit from southern Iraq down to Qatar. He would darken the cockpit as much as possible and do a bit of stargazing. With the night-vision goggles he could clearly see the Milky Way, and each evening, just on that short transit, he would see perhaps half a dozen shooting stars.

'But the difference with the aurora,' Al said, sitting back in his office chair, 'is that it's sustained and it's moving; it's active, it's changing. A shooting star is beautiful but over very quickly. When you see the aurora you can stand there open-mouthed for thirty minutes, just enjoying it. Any light show like that is an amazing thing to see.'

One of the best displays Al has ever seen was from right there in Lossiemouth. Al grew up in Stornoway, in the Outer Hebrides – northerly and dark – so as a child he would see the northern lights quite frequently. In the RAF, whilst training in Goose Bay in Labrador, in the far east of Canada, he also saw some good displays. 'I remember staggering out one evening from the pub and they were just superb,' he recalls. It covered most of the sky. 'But then,' he said, 'last year in Lossie we had some really good coloured ones. Often you just get the green, but this was easily a third of the sky in pinks, greens and red – and this was all visible from in front of my house with the street lights all around.' I smiled at the thought of it.

A delivery arrived, and as Al signed the delivery note through the open window of the portacabin I listened to seagulls calling overhead and the high beeping noise of a vehicle reversing somewhere nearby on the industrial estate. The man departed with a bright 'cheers mate!' and Al turned back to me.

'I suppose one of my favourite views of the aurora is from the International Space Station,' he said. 'That view of it is stunning.'

'Me too!' I agreed. 'It's lovely to be able to see the structure and the changes in colour with altitude. We can see arc-like structure from the ground, of course, but from the International Space Station we can see so much of it. The world stretches out underneath and we get to see the aurora's extent. We get to appreciate that it's not a local thing at all, it's really a global thing.'

'Yes, you can picture the planet actually flying through the storm and . . . well, you can tell me all about the physics!'

Going back to his experiences whilst flying, I asked Al, who mostly flew Tornadoes during his career in the UK, whether he had ever encountered any difficulties due to space weather when he was a pilot. He hadn't, as far as he was aware. 'Nothing that was at the time attributable to space weather,' he told me, 'though there were plenty of times when we had troubles with the Tornado radios . . . they just wouldn't work as well as normal.'

The problem with radio communications is that they work by bouncing signals off the ionosphere. To transmit over longer distances, around the curvature of the Earth, physicists exploited the plasma nature of the upper atmosphere. Plasmas like the ionosphere have a natural frequency of oscillation – the electrons swing back and forth. A build-up of charge in one place creates an electric field that pulls the particles back the other way with such force that they overshoot and build up in an opposite place, perpetuating a cycle that repeats about ten million times a second. This frequency of oscillation is what physicists call the plasma frequency, and ten million times a second is a frequency of ten megahertz (MHz). The ions also move, but the electrons are very small and light compared to the ions, so they respond to the electric field much more rapidly. It is the electrons that are important for reflecting the electromagnetic waves like radio.

In fact, the typical plasma frequency in the ionosphere changes with altitude. It varies between one and ten megahertz as we move through the atmosphere. The electrons oscillate about ten million times a second in the lower ionosphere and this frequency reduces with density – that is, with the number of electrons – the air becoming more and more rarefied the higher we climb. The megahertz range is the High Frequency (HF) radio band, so it is HF radio signals that are most affected by the ionosphere and therefore which are used for radio broadcast.

The relationship between the frequency of the radio wave and the frequency of the plasma it meets determines whether the wave is absorbed, reflected or transmitted. If the frequency of the radio wave matches that of the ionosphere, the radio wave will be absorbed. This is the principle of plasma heating in fusion machines, but for radio transmission around the globe we require the signal to be reflected. For this to happen the radio wave must be vibrating at less than the ionospheric plasma frequency. It is as if anything wiggling more rapidly than the radio wave acts as a mirror. If we want to transmit the signal through the ionosphere to a satellite in orbit around the Earth then the frequency of the radio wave must be higher, and the higher it is above the plasma frequency the straighter the wave will travel.

The way that radio transmission is inextricably linked to the state of the ionosphere means that HF radio communication is highly susceptible to space weather disturbance. It suffers fade-outs and mis-transmission. Early operators were aware of the variable nature of the ionosphere even if they didn't know the cause. Since using a higher-frequency signal resulted in better transmission (up until the 'maximum usable frequency' above which the signal would pass through the ionosphere rather than be reflected), scientists in the 1930s developed models to calculate the best frequency to use. The original model was based on 'normal' ionospheric conditions but it could be adjusted based on observations of solar activity and predictions of ionospheric disturbance. The window of frequencies available for broadcast was constantly changing; operators had to listen carefully to forecasts and predictions. This was the earliest form of space weather forecasting. The Space Weather Prediction Center (SWPC) in Boulder, Colorado, started life as the Radio Propagation Lab.

Space weather affects radio in several ways. Reconnection in the magnetosphere tail accelerates particles into the atmosphere in high-latitude regions – though during big solar storms this ionised auroral region will spread to mid-latitudes – disrupting the ionosphere and causing absorption of HF signals on the night side of the planet. High-energy solar particles entering directly through the polar cusps can cause disruption over the poles. X-rays given off as solar flares can increase the density of the ionosphere on the day side of Earth, resulting in fade-outs. Some communications will also suffer distortion at equatorial latitudes due to electric fields and currents associated with the Dungey Cycle. Fortunately, modern systems can mitigate these effects to some extent, but there remain several older systems in use – many in commercial airlines – that may be interrupted even under moderate space weather conditions.

Al decided that we should taste the Aurora beer. He popped out of the office for a few minutes and returned with two tall, bulb-shaped, narrow-stemmed glasses half-full with the hazy, golden Aurora. Described as 'light on strength but big on flavour', they use Motueka hops from New Zealand to give the beer a distinctive taste. I took a sip. It was light and refreshing, with a slight fruitiness. I say this knowing very little about beer, but it tasted nice.

'We add hops later on after it's fermented, just to really infuse the hoppy flavours into the beer,' said Al as we sipped, 'and New Zealand hops tend to be quite fruity, so the beer itself is very light on the palate and sharp, with a bitterness for a refreshing side.'

'So with the New Zealand influence it's also a bit of a southern aurora,' I said, 'an aurora australis.'

We talked a little of hop varieties, and how different countries grow hops with distinctive characteristics. Like the New Zealand Sauvignon wines, which tend to be quite crisp and zesty, the hops have a citrus-to-tropical kind of flavour. In Britain the hop varieties tend to be more earthy and flowery, making a more flowery beer. German hops have a bit of spiciness to them; American ones are more citrus in flavour.

'Brewing is a balance,' said Al. Brewers have to adjust the flavours of the malt, the yeast and the hops to achieve a clean, interesting beer. 'You can throw a whole bunch of other things into the brew as well, if you like. You can use all these experimental ingredients.' Like any good experiment, taking thorough records all the way through is crucial. And because bitterness receptors are in the back of the mouth, when tasting beer you actually have to drink it, contrary to wine tasting where the wine can be spat out.

As we sipped on Aurora beer, Al phoned the Met Office at RAF Lossiemouth to ask if they had experience of space weather. 'It is becoming a major issue,' said James Fearnley, a forecaster at the Met Office, when we were put through. In addition to radio problems, an extreme storm could affect global navigation satellite systems such as GPS and cause radiation problems for pilots. 'GPS signal paths from GPS satellites will be affected by the heating of the ionosphere, which will affect the accuracy of the GPS systems,' James said. 'Another issue is that anyone flying at high altitude and high latitude during a radiation storm is going to have to be diverted.'

In recent years, with the creation of the Met Office Space Weather Operations Centre, points of contact have been set up at high levels in the RAF so that if they receive a forecast of a big incoming solar storm they can put out information to the groups likely to be affected by it. 'It's a growth area and one we are trying to expand on at the moment,' James told me. 'We're trying to limit the impact – make sure that all the back-up systems are working for locations and GPS and satellite communications. If they do have to divert aircraft, that will be handled at fairly high level and fed down.'

Communication and navigation are vitally important to the military. In wartime particularly, the necessity of communications reaching the desired location – the right receivers – is paramount. Lives depend on it. During the Second World War, in the early days of both aviation and radio, communication and navigation were vital for pilots just as they are today, particularly when flying over empty, monotonous landscapes like deserts or oceans. Erratic space weather could, and did, cause problems, though at the time this was poorly understood; the traditional compass could be affected by the disturbance of the magnetic field that occurs during a solar storm and the new radar technology also suffered interference. In her book *Sentinels of the Sun: Forecasting Space Weather*, Barbara Poppe shares a tale from September 1941 of British bombers failing to return to England from America, their flight home coincident with 'the greatest display of aurora ever seen in Washington, D.C.'. Could the magnetic disturbances have caused compasses and radio communications to fail? There is another, similar, story from February 1944. A British plane disappeared without trace, assumed lost over the North Sea or the Atlantic, during a time of high geomagnetic storms.

Nowadays, aircraft, shipping and even the general public use satellite navigation, no longer having to rely on a magnetic compass. Satellite navigation such as GPS has simplified and improved navigation, but it is not infallible. GPS works by deploying the old triangulation method used by the Great Trigonometric Survey to measure the heights of peaks in the Himalaya, and which Størmer later used to calculate the height of the aurora from separate, synchronised photographs. These applications used position and angle information to calculate a distance (the height), whereas GPS uses multiple distances to calculate position.

A GPS device, such as your phone or a navigation unit in an aircraft, receives a signal containing position and time-of-broadcast information from four satellites in view. It notes the exact arrival time and compares it with the broadcast time to calculate how long the signal was travelling from satellite to receiver. Knowing the speed of light, the GPS device can calculate its distance from the satellite. Using the multiple satellites allows the device to pinpoint its exact location with reasonable accuracy.

Travelling from the satellite to the device on the ground, the signal passes through the ionosphere. Though the signal is only passing through and not bouncing off as for HF radio communications, a disturbed ionosphere can still cause problems. If there are extra electrons being fired down the geomagnetic field lines into the ionosphere during reconnection events, this will change its density. There may be patches of higher density where electrons cluster together; in these patches the GPS signal will be slowed down, resulting in an inaccurate calculation of distance at the receiver. A longer transmission time will seem as if the signal has travelled from further away. Another problem is that larger numbers of charged particles in the ionosphere can cause scintillation of the communication signal, which is like the stars twinkling in the night. Whilst this may be a pretty effect for faraway stars, scintillation of electromagnetic waves being used for communication causes the signal to fade in and out and transmission to fail.

The aurora itself may be able to help predict where GPS scintillation and consequent failures may occur. There is no direct physical link between the visible aurora and GPS scintillation – that is, the light emission does not cause the problem – but the light emission of the aurora comes from a region of intense ionisation in the ionosphere triggered by electrons pouring into the area, conditions that are known to cause scintillation. Simply viewing the aurora can give a picture of the small-scale structure of the ionosphere, which can range in scale from hundreds of metres to tens of kilometres. The aurora may thus indicate GPS scintillation risk, particularly during intense solar storms where defined auroral arcs are present. Ultimately, though, GPS may be rendered inaccurate, unreliable or unusable during large solar storms.

If satellite navigation systems have issues with space weather, satellite communications will suffer similarly. Civilian communication systems should only have problems during intense storms since they operate at a higher frequency than the VHF and low-UHF bands used by the military. Satellite broadcasting operates at much higher frequencies again – around ten gigahertz (GHz) – so is unlikely to be affected provided the broadcasting satellite remains operational.

After the Second World War and with the advent of the space age, through the 1960s and 70s satellite communications services established themselves, distributing telephone and television signals. On 20th July 1969, over 600 million people watched the Moon landing on television, broadcast through a global network of Intelsat satellites. Around the same time, in the 60s, commercial aviation was booming. Passengers had been travelling by air since before the war, but only in small planes and in low numbers. Larger aeroplanes designed in the 1950s increased capacity, and the invention of the jet engine in the early 1960s increased speed and efficiency.

When satellite communications superseded radio in aircraft, the earlier HF radio system was relegated to a back-up. The new system used Very High Frequency (VHF) signals bounced up to satellites in geostationary orbit, hovering over the same point on the equator, therefore not relying on transmitters on the ground. However, using satellite communication is impossible when travelling over the poles, meaning that aircraft must switch to the back-up HF radio whilst travelling over the regions of the globe most vulnerable to ionospheric disturbance.

There are advantages to travelling over the poles and thus cost savings to be made for airline operators. Aeroplanes travel shorter distances and avoid the strong jet streams encountered at lower latitudes. Several popular itineraries, such as London to Hong Kong or Chicago to Singapore, take such a route over the North Pole. But the eight degrees of latitude polewards of 82° cannot be reached by the satellite signals broadcast from the Equator. Aircraft here are in the shadow of the Earth itself, hidden behind the curvature. If the ionosphere is disturbed by a solar storm and the HF radio is rendered useless too, aircraft in this region are exposed to an unacceptable outage in communications. Disturbances can range from minutes to days – some large storms have caused outages of up to 120 hours! Aircraft simply cannot fly over the poles in these conditions, and so in such situations airlines have no option but to reroute flights, sometimes at a cost of hundreds of thousands of dollars per flight to cover the extra time, fuel and staff.

Flights over the poles during periods of adverse space weather may be unacceptable for another reason too. The increased bombardment by energetic charged particles raises the radiation exposure of crew and passengers, which may have a long-term effect on health.

According to Wade Allison in *Radiation and Reason*, radiation is 'energy on the move'. It can be either a wave or a particle, like a ray of sunlight or a fast-moving neutron. The whole of the electromagnetic spectrum is radiation of varying energy, ranging from the long-wavelength, low-energy radio waves through microwaves and visible light to the short-wavelength, high-energy X-rays and gamma rays. Alpha radiation is fast-moving helium nuclei; beta radiation, like that used in PET scans, is electrons and their anti-matter pair, positrons. Some radiation is benign and we call it non-ionising. This is the low-energy radiation such as visible light or radio waves. Other types, those of higher energy, we call ionising because they are capable of breaking molecules apart. It is with this type of radiation – the ionising radiation – that we need to be careful. This radiation can cause cell damage or DNA mutation if received in elevated doses.

Radiation is naturally present in the environment, occurring in water, soil and rocks such as granite, as well as being taken up by plants we eat, and it is present in our own bodies. Almost ten per cent of our natural, annual radiation exposure is from the decay of radioactive atoms within ourselves. Humans and other animals have evolved in this environment and so have natural repair mechanisms to deal with low levels of radiation. Antioxidants limit the early, chemical changes in cells caused by radiation, certain enzymes repair minor DNA damage that may occur and, if damage is too much, a cell suicide process called apoptosis* is triggered to ensure replacement of dead or damaged cells.

---

* The word 'apoptosis' comes from the Greek for 'falling off', as in leaves falling from a tree.

About twelve per cent of our annual radiation exposure originates from space and we call it cosmic radiation. This is mostly muons – a heavy kind of electron – created by collisions of fast protons with the atmosphere. As well as being emitted by the Sun, high-energy protons can also come from outside the solar system, and even outside the galaxy. The magnetosphere and the atmosphere protect us from the majority of incoming charged particles. As we now know, charged particles cannot penetrate the Earth's magnetic field, but they can enter via reconnection events, and electrons within the Earth system may be accelerated into the atmosphere and cause the aurora. This radiation does not reach the ground and has no impact on human health; however, problems arise when there are higher-energy incoming particles and less atmospheric protection, such as when one is flying.

At sea level we receive an annual dose of 2.7 millisieverts from natural radiation. The sievert (Sv) measures the combined effect of various types of radiation so that the probability of long-term effects such as cancer may be gauged. One sievert is actually quite a large dose, so we usually talk of millisieverts (mSv), or thousandths of sieverts. Allison cites evidence from studies of cancers and leukaemia in nearly 100,000 Hiroshima and Nagasaki survivors over a fifty-year period that suggests that a single dose of below one hundred millisieverts has no effect on life expectancy. Yet a recent study published in *The Lancet* and reported in *Nature* indicates that long-term exposure to low-dose radiation does indeed increase the risk of leukaemia, although the rise is thought to be minuscule.

Long-haul air crew – the most highly exposed occupation – receive around four to six millisieverts per year. NASA says that flight crews on high-altitude routes have a higher radiation exposure than nuclear plant workers. However, radiation limits are conservative and this is a low dose rate. There is currently no evidence to suggest that air crew are at a greater risk of cancers from this radiation exposure, although they do seem to be more at risk of skin cancer and breast cancer than the rest of the population, possibly due to the disruption of circadian rhythms suffered as a result of overnight flights and shifting time zones.

However, whilst ordinary conditions may not cause a problem, solar storms generating high-energy particles enhance the level of ionising radiation at the altitudes at which aircraft fly. Some particle events are so intense that the high-energy particles can be detected at ground level. They are referred to as ground-level events for this reason. Radiation levels can be increased by a factor of a thousand at cruising altitudes, so air crew may receive a dose of several millisieverts in an hour, meaning that over a single flight their annual radiation limit could be exceeded. (This dose is still well under the one hundred millisievert level below which there is limited evidence of change in life expectancy, but above legal limits.)

Flights over the poles are more problematic again, since there is more exposure to solar radiation due to the polar cusps, the openings in the Earth's protective magnetosphere. The effect of increasing latitude is actually greater than that of increasing altitude. Supersonic, high-altitude routes such as those flown by Concorde in its operating years are less exposed than subsonic (low-altitude) high-latitude routes such as those that ordinary commercial flights take over the poles. Concorde was obliged to carry a radiation-warning monitor, but there is currently no such requirement on subsonic airlines. Currently it is almost impossible to forecast high-energy particle events because the particles travel at close to the speed of light and so are already with us when a flare is seen. Better aircraft monitoring would at least allow air crew and passengers to monitor their radiation exposure and risk, particularly after encountering a solar storm.

Clearly airline operators need to be aware of solar activity and the space weather conditions in the near-Earth environment – both actual and forecast. Action plans should be established for radiation increases. Such large occurrences as the ground-level events are, fortunately, rare. Since 1942 there have only been six events that would have caused legal radiation limits to be exceeded over a representative long-haul flight from London to Los Angeles. However, alongside a desire to protect air crew, there is also a need to minimise overreaction, particularly relating to measurements of lower-energy particles that may affect satellites but will have little, if any, effect on aircraft. As seen during the volcanic eruption of Eyjafjallajökull in 2010, grounding of flights can cause unwelcome social and economic chaos, and fear of radiation exposure can often cause psychological effects more serious than the actual radiation damage.

If charged particles can cause this much trouble within the protective atmosphere of the Earth, what of satellites and astronauts?

The hazard to astronauts is, of course, greater than for airline passengers because they are at higher altitude. Astronauts need to be warned of coming radiation storms so that they can take cover in more shielded parts of their spacecraft. Generally, astronauts in space remain within the Earth's protective magnetosphere, which shields them from much of the incoming radiation, but the magnetosphere is dynamic and its extent can change, so not all spacecraft are shielded all the time. Leaving the spacecraft on a space walk also intensifies the risk. Spacesuits offer some protection, but with current technology it is difficult to design protective suits that are also manoeuvrable and allow vision. Going outside of the Earth's environment would bring even greater health problems, so radiation protection and safety is of critical importance for a manned mission to Mars (being planned by NASA for the 2030s). Ruth Bamford from the Rutherford Appleton Laboratory has proposed an artificial mini-magnetosphere for the spacecraft that would deflect charged particles away in the same way that the Earth's magnetosphere does.

Besides astronaut safety, radiation damages satellites and other equipment. NASA will not launch on days when there is a high risk of a geomagnetic storm because they want the best chance of a successful launch. Minor equipment failure can be very costly in such a situation. Fortunately for NASA, there is not too much of a space weather problem at Cape Canaveral in Florida, but launches have been cancelled from the Kodiak Launch Complex in Alaska because of strong geomagnetic activity.

For a satellite in orbit, all the charged particles flying around can affect not only its operation but also its lifetime. Careful design of spacecraft and choice of orbit can improve its chances, but can't protect it from every eventuality. A solar storm can dramatically enhance the number and speed of the particles that the satellite will encounter, and the more extreme the storm the greater the potential for damage.

The main particles likely to hit a satellite are electrons and protons. Electrons are trapped in the magnetosphere where the satellites are orbiting; electrons, protons and also some ions are accelerated from the Sun during solar flares and coronal mass ejections; and some high-energy protons and ions even come from outside the solar system.

The Earth's radiation belts are where there are large collections of electrons bouncing back and forth from north to south, trapped by the geomagnetic field. The varying orbits of satellites take them through different regions of the radiation belts, but few seem to avoid them entirely. The geostationary satellites hit the outer edge, and those in medium Earth orbit fly straight through the centre of the outer belt. Satellites in low Earth orbit generally pass underneath the belts, although those with steeply inclined orbits that traverse high latitudes will meet the end of the outer belt. The trajectory over the poles also leaves them exposed to protons that pass into the Earth's environment through the openings of the magnetosphere.

The solar energetic particles (SEPs), those accelerated from the Sun with flares and coronal mass ejections, come in two main streams. Those associated with flares are dominated by electrons, along with some ions like helium and heavy elements like iron.

They are released suddenly in bursts. Those that are accelerated in shock fronts produced by coronal mass ejections are mostly protons, and the acceleration occurs more gradually, but it is these events that produce the most solar energetic particles. The bursts from flares come from a single point on the Sun, whereas the shock wave can spread out laterally so it encompasses a wider area and accelerates more particles. It is the high-energy protons produced in this way that are the significant radiation danger to astronauts, air crew and equipment.

Thus the satellite environment is complex and hostile. Even within the confines of the magnetosphere there is still some penetration of solar energetic particles, and the energies and extent of the radiation belts can change with the strength of solar storms. Storm conditions pump energy and particles into the magnetosphere and previously safe satellites may find themselves passing through much higher particle concentrations than expected. But what is it that these particles do to the satellites? The main difficulties are charging, upsets and ageing.

Charging happens when electrons hit a satellite (either the surface or internal structures, depending on the energy of the electron) and build up in a particular place. Electrons have a charge, so accumulating electrons means accumulating charge. If the charge builds up differently in different places then eventually a current will flow between them to even everything out. This is called an electrostatic discharge. It's a bit like miniature-scale lightning, which equalises a build-up of charge in the clouds by discharging down to Earth. Alas, discharges often occur close to sensitive components.

Protons that hit the satellite can disrupt electronics. The protons are bigger so they can knock atoms out of position in the organised structure of microelectronics, also depositing charge that can cause some components like semiconductors to burn out. As electronics get more sophisticated and smaller there is greater scope for damage. Satellite operators call these small disruptions to the electronics 'upsets' and too many can lead to operational outages and failures. There are many more upsets than usual during an intense solar storm.

Fortunately, service outages and satellite failures are rare, despite the hostile space through which satellites travel, though of course these problems increase with storm severity. During the 2003 Halloween storm, which lasted for several days, around fifty satellites reported problems and ten were out of operation for over a day – one for as long as two weeks! One scientific satellite, the second Advance Earth Observing Satellite, ADEOS II, was lost entirely when its solar panels failed. This prematurely ended its mission only ten months after launch.

Finally, being hit by fast, charged particles, particularly being hit hard by fast-moving ones with lots of energy, causes cumulative damage to the satellites and may lead to premature ageing. Electrons and protons contribute to the ageing of satellites. As well as doing their utmost to protect their fleet, operators also need to have an awareness of changes in the particle environment due to large storms that may affect satellite longevity. In this way they can ensure replacement of critical satellites before it's too late.

Radiation issues with satellites can have knock-on effects in other sectors that depend on them. For example, satellite navigation systems and satellite communication, already affected by space weather due to signal transmission problems, could be rendered totally inoperable if an upset should cause an outage or failure. Alternative options are a necessity. Terrestrial broadcasting could also see problems if satellite power were to fail or GPS timing lost accuracy. It may not be life and death, but even television operators are providing a service and they do not want to let their customers down. Small particle-induced upsets have the potential to cost satellite operators dearly.

Towards the end of my visit to Windswept, as Al was showing me around the brewery, two large men came in wearing khaki

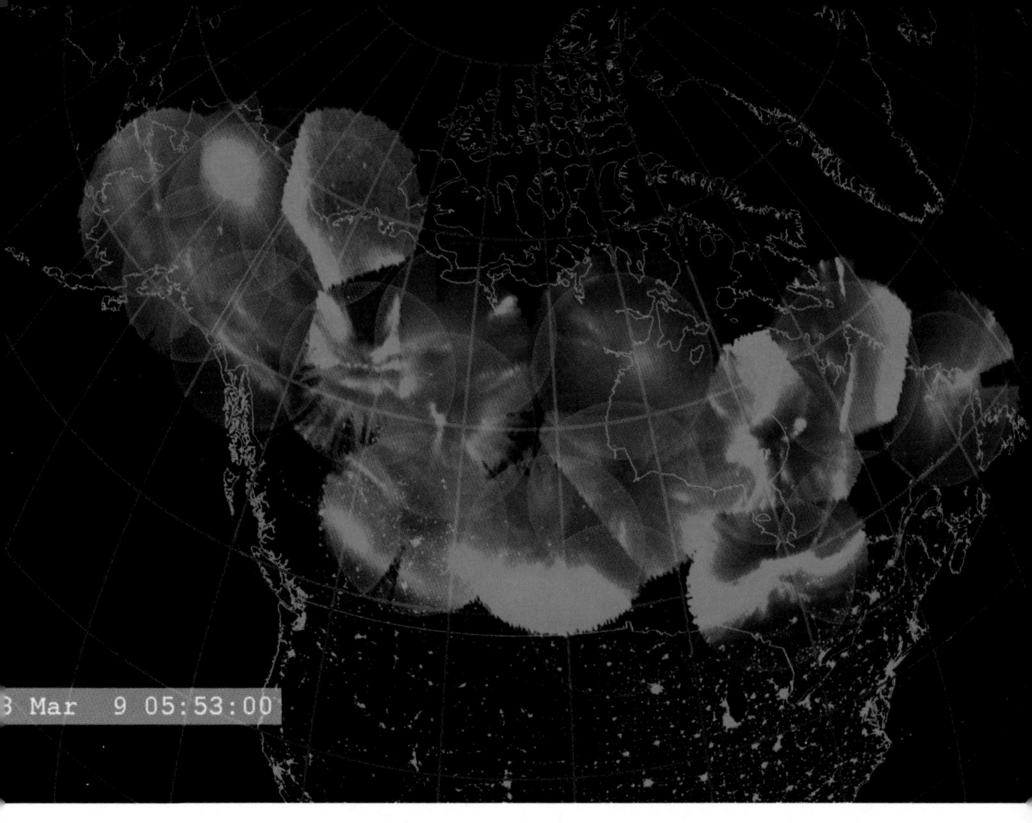

Mosaic of aurora from the THEMIS All-sky Imagers showing aurora stretching all across the North American continent. © NASA/Goddard Space Flight Center Scientific Visualization Studio

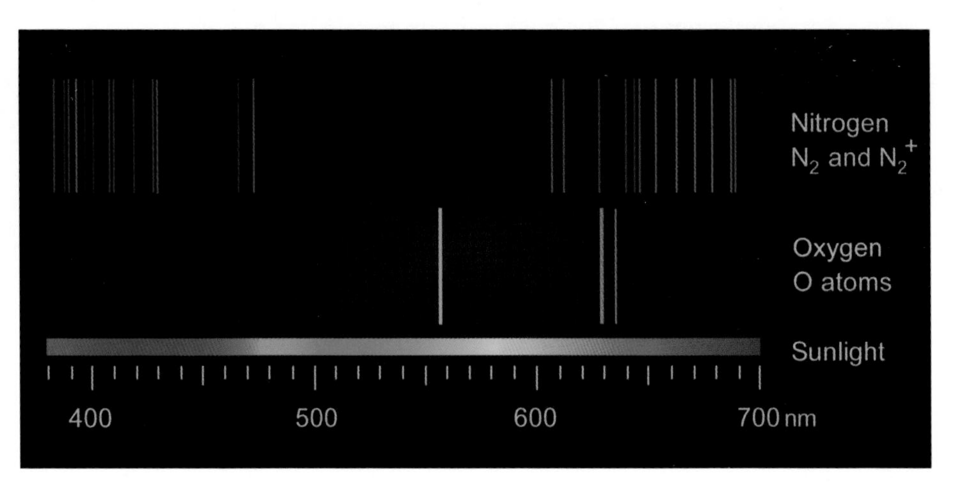

The individual colours of the aurora. These are the spectral lines of the aurora, showing which colours come from oxygen and which from nitrogen. Early spectroscopists did not recognise these colours as coming from atoms they knew on Earth. They thought the colours came from new, undiscovered atoms. Instead, they are unusual, 'forbidden' transitions of common elements that only happen in near-vacuum conditions. © Les Cowley (atoptics.co.uk)

A bright green aurora over a lake near Yellowknife, Northwest Territories, Canada. Being directly under the auroral oval, aurora are seen almost every clear, dark night in Yellowknife. The dominant colour is green. © James Pugsley, Astronomy North Society

An AuroraMAX image from Yellowknife, Canada. I love the rayed structure of the aurora in this picture. The colours seem to rain down. © AuroraMAX

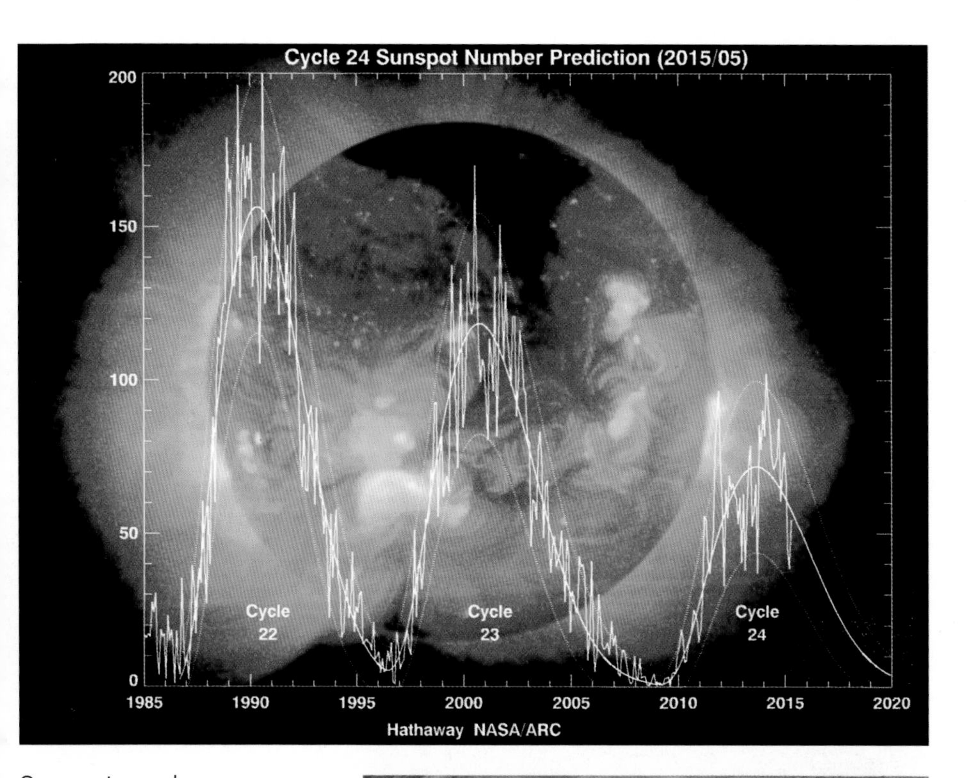

Sunspot numbers over the last three solar cycles. The most recent cycle has been quieter than usual and NASA has termed this solar maximum a 'Mini-Max'.

A painting of the Duncansby Stacks in summer by David Cameron Watt, Scotland.

My best ever photograph of the aurora (so far!). Caithness, Scotland.

Looking out at the aurora from Castlehill Heritage Centre in Caithness, Scotland. The intensity of this aurora was very low, so to the naked eye it looked more like a white sea fog. © Chris Sinclair

Me skiing along Reindalen, Spitsbergen, Svalbard. © Svante Strand

(above) A star pattern on the floor of Seville Cathedral. (right) The 'global view' showing the floor pattern is actually squares. © Jesper Gjerloev

Composition of images showing the solar corona during the Svalbard eclipse. We can get an impression of the Sun's magnetic field from the lines in the corona. © Miloslav Druckmüller, Shadia Habbal, Peter Aniol, Pavel Štarha

The diamond ring effect and pink solar prominences seen on the Sun during the Svalbard eclipse.

The eclipse seen above Svalbard, 20th March 2015.

green flight suits and heavy black boots. Their suits had large, bulging pockets, even low down on the legs, and on the shoulders and arms were thick lapels and several badges and patches. Their hair was cropped short.

'Alright!' said one of the men, smiling broadly.

'How's things, guys?' said Al.

'It's good, how're you doing?'

'Yeah, good thanks,' said Al, shaking their hands one by one. 'Canadians?' he asked looking at their flight suits.

'Yeah. One of our guys came down here yesterday. Says you do some good beer.'

Al grinned. 'Whereabouts in Canada are you from?'

'Nova Scotia.'

'Oh, well you guys will be really familiar with the northern lights then.'

'Yeah' they said disinterestedly.

'I'm writing a book about the northern lights,' I explained.

'Oh, you're writing a book!' One of them said, perking up a little.

'Yes, that's why I'm here visiting.'

'Oh wow, look at that. Well, y'know what, uh, we used to have a patch with the aurora on it.' He turned and showed me his left arm where there was a large, circular patch, about 8 centimetres (3 inches) across with an aeroplane flying out from the centre, wings stretched across the diameter, and a large, red maple leaf in the background. Top and tailing the plane around the circle were the words 'CANADA AURORA'.

'We used to have a really cool crew patch that actually had the aurora borealis on it,' he continued. 'Back when we were allowed to wear colour.'

'So are you guys on the P3 Aurora?' asked Al.

'Yeah. We don't call it a P3, though, we just say, "Aurora".'

'You're flying an aurora aeroplane?' I asked excitedly but semi-incredulously.

'We couldn't have laid this on any better, could we?' said Al.

'Look, I tell you what,' said the pilot, 'you gotta have this patch!'

* * *

WHEN THINKING ABOUT civil contingencies, something that captures peoples' attention and imagination more than any other is the vulnerability of the power transmission networks. It is one of the main reasons why space weather is on the National Risk Register. Electricity is now so firmly embedded in our society that a loss of power for any amount of time has considerable impacts, cascading operational failures down through a chain of linked industries. Businesses would suffer initially but, soon, so would life in general. Water distribution would be affected within a few hours; with a lack of refrigeration perishable food would be spoiling in twenty-four hours; depending on the temperature, people would be suffering from the lack of heating or air-conditioning in their homes; fuel and transport would be affected because electricity is required for pumping; communications would fail; sewage works would be down without power or water; hospitals would be struggling with medicines that need refrigerating and back-up generators would soon run out of fuel. These are just some of the ways in which society could be affected by a long blackout.

In August 2003, some overgrown trees interfering with sagging power lines in Ohio tripped power failures across the north-eastern United States and into Canada. The blackout – the largest in North American history – affected 50 million people over two days and cost an estimated $6 billion.

Realisation of the effects of space weather on power transmission systems came later than for radio simply because the technology developed later. Yes, there were small fires and electrical disturbances on the telegraph, and power had been transmitted locally in various experimental ways since the end of the 1880s, but the idea of a national electricity network like the UK National

Grid only came into existence in the 1930s. In the UK it began as regional grids constructed at the end of the 1920s. The Central Electricity Board was created in 1926 with the mission to link power stations with consumers via a single, interconnected system. The first electricity pylon in the UK was built in 1928 near Edinburgh, and the regional grid in Central Scotland was the first part to open, in April 1930. By 1935 there were seven operating areas across the UK, forming the first integrated national grid in the world. From the beginning of the Second World War the entire National Grid was operated as a single unit from London.

To find out more about how space weather can affect the electricity grids I spoke to Andrew Richards, Severe Risk and Resilience Analyst at National Grid in the UK. I asked him when people first began seeing space weather having an effect on power distribution.

'1940 is the first time we know of any disturbances on an electricity grid, in the sense that we have them today, that can be attributed to space weather,' Andrew told me. 'On Easter Sunday in the north-eastern US they had some big voltage problems. There were no blackouts and no damage, but they wondered what on earth was causing it. At the time it was attributed to solar flare activity on the Sun. But the big bit of key evidence – in my world – is 1989.'

In March 1989 there was a big solar flare and a coronal mass ejection – in fact, several coronal mass ejections one after the other – that hit Earth. The resulting big geomagnetic storm hit most strongly in Canada over the Québec province. Overnight the electricity grid operators at Hydro-Québec experienced difficulties controlling their voltage. They had no idea what was causing it but they thought they had it under control until, suddenly, a couple of lines connecting their big hydro generators in the north to their population centres in the south tripped. The instability that the trip created in the network caused an electrical blackout for about ninety per cent of Québec province. The voltage on the transmission lines collapsed within about ninety seconds, during which time the operators could do little but watch until the lights went out.

It took Hydro-Québec almost ten hours to restore power. The reason they were able to do it so quickly (yes, ten hours is quick for this kind of job, according to Andrew) is because they were able to reboot the system from the north-eastern United States. It was lucky that their grids were connected; for an isolated system like the UK grid, restoring power would take much longer. The estimated economic cost of the blackout is over ten billion Canadian dollars. Even now Hydro-Québec is quite touchy about the subject and doesn't want to talk publicly about it. But they, and others, have learnt from the experience.

'It is that 1989 event that first made people really sit up and think "this is space weather",' said Andrew. They ascribed it to coronal mass ejections coming from the Sun. 'That's the first time that we as an operator started to think that we must take this seriously,' he continued. 'We started to plan our operational response to space weather from that moment.'

To understand fully the effect of space weather on the grid, in particular the geomagnetic storms that cause the aurora, we need to understand the nature of the electrical system in the UK. This comprises generation, transmission and distribution, but the National Grid is just responsible for the transmission part, the transportation of electricity around the country. This is the part of the system that is vulnerable to space weather effects.

There is a range of electricity generation across the country – from the large electricity generators running power plants on coal or gas or nuclear, to the smaller generators that perhaps operate a small hydro plant or a solar farm, and right down to solar panels on people's houses. The electricity that is generated needs to be transported to where it can be used. The National Grid has about 8000 kilometres (5000 miles) of transmission lines and cables on huge pylons stretched across England, Wales and Scotland like motorways of electricity. Then distribution is the very small-scale transportation of electricity to the end user. Part of the distribution network is the electricity wire that goes into your house, assuming you are connected to the grid. The different networks use different voltages in the wires depending on how far the electricity has to travel and how far away the cables are from town centres and people.

Electricity is the movement of charge, performed by negatively charged electrons. Electrical conductors have 'free' electrons which are only loosely bound to their atoms, so they can move around if acted on by an outside force. Voltage provides this driving force; it is simply a discrepancy in charge between two points, known as potential difference. Just as happens in a plasma, a build-up of electrons (charge) in one place over another causes an electric field to be set up that acts to move the electrons. This flow of electrons is what we call the current. Batteries use a chemical reaction to cause a net build-up of charge at the terminals (a potential difference) which drives current around a simple circuit. To get a potential difference across the distance of a National Grid transmission line, so that a current will flow, requires both a generator and a demand, which is something using the electricity at the consumer end. The grid is a complex balancing act – supply must equal demand – but voltage can be manipulated between the end points, and it is beneficial to do this to reduce power losses in transmission.

Wires have resistance. This is a bit like a friction that impedes the flow of the electrons. It arises because the electrons collide with other electrons and fixed atoms in the conductor as they travel. During these collisions the electrons lose energy, transferring it to the atoms which vibrate faster, increasing the temperature of the conductor. So electrical energy becomes heat. The higher the resistance the more electrical energy is lost as heat. Also, the more electrons are flowing; that is, the higher the current, the more energy is wasted in this way. So if you're going to send lots of power somewhere, you need to do it at low current and you need to use very low-resistance wires to minimise losses. These conditions correspond to high voltages.

The wires that the National Grid uses for the transmission network are specifically built for the purpose and are very expensive. The voltages used for transmission are 400,000 volts or 275,000 volts. Compare that to the 240 volts going into your house or the 12 volts of your car battery and it is immediately clear that these are very high voltages, which is why they are kept away from people. If you got too close to a high-voltage wire when you were standing on the ground a very big current could jump across the air gap and pass through you to the ground, which wouldn't be very nice (in fact, it would most likely be fatal). From transmission into the distribution network, closer to people, the voltages are dropped down to 132,000 volts, then to about 3000 volts and finally to 240 volts as the electricity enters homes.

These conversions of voltage are made by huge transformers, which use the properties of electricity and magnetism to step up and step down voltage and current in inverse proportion. Generators have transformers that step up their voltage to connect to the Grid; National Grid owns the transformers that step down to the distribution network; and the distribution companies own transformers to step down further from the voltage they receive from the Grid to that which goes to the final consumers.

Voltage is a really tricky thing, made even more complicated by the fact that the electricity network is operated with alternating current. A revolving turbine generates electricity in cycles, with current and voltage varying sinusoidally up and down fifty times per second (50 Hertz – or 60 Hertz in America). As well as this inherent oscillation, voltage can fluctuate if the supply and demand get out of balance or depending on the types of load that are put on the system. For example, an appliance that rotates, like a washing machine, has a different effect on the Grid to something static like a light bulb. If the voltage fluctuates too far from where it should be, the fluctuation can be exacerbated by the generators themselves beginning to speed up and slow down in delayed compensation. Eventually the unstable voltage swings down into the danger region and the voltage collapses to zero. Once the instability sets in there is nothing the operators can do about it, though it is possible to isolate the problem in a particular region by flicking switches to cut off the troubled part of the Grid.

Voltage fluctuations that result in a voltage collapse can cause damage to the large, expensive transformers. Within the affected region of the country there is now no power. There is no demand to balance the supply because everything has gone dead. However, immediately after the voltage collapse the power station may still be generating; the great heavy turbine still spinning and producing electricity which flows into the transformer. But now there isn't anywhere for the current to go because the other side is essentially disconnected, so it just goes through the transformer to ground, heating up the transformer as it goes. The transformer is not designed to take that much current so it just gets destroyed. This very thing happened in Québec in 1989. When the power lines tripped, the hydro generators kept on generating even though the power had nowhere to go. Two transformers were damaged because current continued to flow in. The cause of the damage on that occasion was the space weather storm, but the transformers would have suffered exactly the same effect if a tree had blown across the line.

During that same March 1989 storm, the National Grid in the UK saw two transformers start to misbehave, not because the Grid had fully lost power due to a voltage collapse but because the aurora was affecting current in the transformers. Consequently they were overheating, and when they overheat their insulating oil starts to give off gases. Monitoring sensors detected the gases and the transformers were switched off. They were relatively new so the National Grid suspected they were faulty. (Remember that at this point nobody knew that the disturbance was to do with space weather.) The transformers were sent back to the manufacturers, who checked them and returned them saying there was nothing wrong with them at all. They were plugged back into the system and they are still operating very happily today.

So how does space weather affect the electricity transmission networks and why does it not affect distribution?

During a geomagnetic storm, the Earth's magnetic field varies rapidly. It is buffeted by the solar wind and fast plasma from coronal mass ejections, and within the auroral oval currents flow in the ionosphere and cause further fluctuations on timescales ranging from a few seconds to tens of minutes. As the auroral oval expands and contracts with the breath-like movements of the magnetosphere over the substorm Dungey Cycle, the magnetic field across the landmass below changes. To exacerbate the situation, the auroral oval doesn't expand smoothly but rather has kinks and twists that can increase the magnetic variation.

As we first found out in Iceland, changing magnetic fields create electric fields and induce currents to flow in nearby conductors. In the case of the changing geomagnetic field, the electric field created is in the Earth's surface. It's the rate of change of that magnetic field measured at ground level that causes the trouble.

The faster the field varies, the more electric current is induced in the land. The effects of this depend on the geology. Problems occur for the National Grid if the rock over which the auroral oval is passing (and thus the magnetic field is varying) is resistive to the flow of current. If the resistance is high, it is much easier for the current to shoot out of some earthing point, flow along a low-resistance power transmission line and then return to Earth to complete the circuit. If the rock is very conductive, the current just flows through the land and everything is okay.

Current will only flow from the earth into an external conductor if the resistance along that conductor is lower than the resistance of the earth. Similar currents can also flow in long metal pipelines such as the Trans-Alaska Pipeline, which can cause corrosion damage. But it is this propensity of current to always take the easiest route – literally the path of least resistance – that keeps the effects of space weather on the Grid confined to the transmission lines. High-voltage lines have very low resistance, whilst low-voltage lines do not. It is not favourable for the induced current to travel down the low-voltage lines of the distribution network, so this part of the grid – and our homes – sees no effect. The problems occur at the earthing points of the transmission lines, at the high-voltage transformers, when this new current passing through is added to the ordinary current.

The geomagnetically induced currents look almost like steady, direct current alongside the rapidly oscillating alternating current of the Grid (wiggling around at 50 cycles per second). So the induced current shifts the entire baseline of the alternating current upwards. If it pushes above the limits of the transformer design the transformer heats up. Just as happened to the two UK transformers during the 1989 storm, the insulating oil will get hot. If heating continues the oil can boil, setting fire to further insulation around the transformer, which is often in the form of paper or cardboard. In the worst cases the metal windings of the transformer (the wires that wind around the core to transform the voltage up or down) can heat up so much that they can melt.

Fortunately this doesn't happen often. Andrew said that there are only about five instances that he knows of where this has occurred anywhere in the world. But even lesser heating can cause damage that can build up unnoticed and cause failure later. In October and November 2003, there was a storm period that NASA has described as 'one of the largest outbreaks of solar activity in recent history'. After this event, now known as the Halloween Storms, South Africa suffered eight transformer failures – one or two within a couple of days, the others within the year. South Africa is at the lower edge of the mid-latitude region, its northern tip crossing the Tropic of Capricorn, so it is not somewhere typically associated with auroral or space weather activity. However, during the large Halloween Storms aurorae were seen in mid- to low-latitude locations like Texas, Florida and the Mediterranean countries.

Of course, if the transformers weren't earthed then there would be absolutely no trouble at all. But then they would be far too dangerous. If people walked anywhere within about 100 metres (330 feet) of the lines they would be electrocuted to death, so safety requires that they are earthed.

As well as heating, geomagnetically induced currents in a transformer cause it to consume too much power. This can trigger voltage instability and perhaps full voltage collapse and blackout. A full blackout would, of course, affect the general public even though the physical effects don't enter the distribution network. But it would not be obvious that the blackout is due to space weather – it would just look like a blackout for any other reason, like a tree across a line or failure of some equipment. If the voltage collapse has been isolated to a particular region of the Grid due to swift action by the operators, then the people should not be without power for too long. However, if operators don't take action to stabilise or isolate the voltage instability then we could see a national rather than a local blackout.

A national blackout in the UK is much harder to restart than a localised one because, unlike Québec, we don't have a neighbouring big electrical system that we can call on. We would have to start it again from within the country, which is actually quite difficult because you can't just start up an electricity generator unless you have exactly the right balancing demand for what it will be generating.

Firstly, power stations themselves consume power – small amounts of power to make large amounts – so if there is no power, where do you get the power from to run the power station? The National Grid has special stations called black-start stations where a small diesel generator is used to start the turbine spinning and get the power station generating again. At the same time, ready and waiting, there must be a demand to connect to the newly generated power so the current has somewhere to go. Once the first of the power stations is working again it can be used to start the next power station, and the next, so gradually the Grid returns to its fine balance and life returns to normal. But this process of resynchronising the whole system and getting the power up can take between two and three days, during which time the country would have no power.

Although very rare – there has never been a national blackout on the UK network due to space weather – an event of this scale would be a serious national emergency. Not only would it be enormously expensive, but during that time, as a result of power blackouts, some people would die, as occurred during the New York blackout in 2003.

* * *

WITH A BETTER understanding of the workings of the National Grid and the ways in which space weather affects it, I asked Andrew about the potential severity of space weather effects on the UK National Grid, and particularly in Scotland. I wanted to know if Scotland was more at risk than the south due to its higher latitude.

Andrew told me that it is very rare to see any effects in Britain, but the potential effects of an extreme storm are so serious that operators have to be vigilant. 'The last time that we detected any disturbance on the system at all because of space weather, in terms of voltage fluctuations or anything that has to be managed, was in 1991.'

As a result, all the engineers in the control room, those who actually operate the system, don't believe that space weather ever does anything. They have never seen it. Satellite operation managers sometimes have the same issue. They put safeguards in place, develop mitigation strategies and training for an extreme storm eventuality but, as Andrew says, 'I know what the engineers say when I shut the door and leave!'

As for the risk of blackout or damage in different regions, there are so many different effects combined together that it is very hard to say that one area is more at risk than another. Latitude, geology and network topology all come into play. The risk increases further north because there is more chance of the auroral oval expanding into the more northerly mid-latitudes. The geology, particularly the conductivity of the earth, is critically important. This is related to the resistance of the current path and determines whether the current will happily flow through the earth or will seek a less-resistive route through external conductors.

With the help of the British Geological Survey in Edinburgh, the National Grid makes very detailed models of the geology of the UK down to a depth of about a kilometre (0.6 miles), mapping all the different types of rock and their different conductivities. Below that they have a simpler, one-dimensional model of the resistance of the rock down to about 800 kilometres (500 miles), the depth to which the induced electric field penetrates. They can use the models to predict where in the UK currents will flow under varying storm conditions.

Additionally, the topology of the network matters; in other words, how the wires of the network are actually linked together. The UK grid is very connected, like a patternless spiderweb of multiply connected nodes. In contrast, for example, the US grid looks more like a constellation – single nodes joining to only one or two other nodes as the lines span the wide continent and couple isolated power stations to communities. The highly connected nature of the UK network makes it more robust to geomagnetically induced currents. At a node with numerous high-voltage lines there is somewhere for the current to go rather than to ground via the transformer. Based on the topology of the network, rather than any other factors, isolated nodes are much more vulnerable.

The interplay of these different effects means that it is very difficult to say whether some regions are more at risk than others. Scotland, for example, is more at risk than the rest of the country because it is further north and because the granite that is underneath much of Scotland is less conductive (that is, more resistive) than other types of rock.

However, compared to England, Scotland has very few of the high-voltage lines because there are far fewer large conglomerations of population. Instead, there are many small towns and villages. The transmission lines of highest voltage run from Edinburgh to Glasgow, one right to the north of Scotland, and there are two running south down into England. The lower voltages reduce the chances of the lines being a favourable path and give the Scottish part of the system a bit more robustness despite its northerly location. Yet, with all these competing factors, the final risk is very difficult to assess.

From all the modelling work that Andrew has done, examining what happens across the country in different scenarios, he believes that there is almost zero chance of something like a months-long, UK-wide blackout, though people talk about this possibility in the US. This low chance is due to contingencies and back-ups that are in place.

'We have models that tell us how likely a transformer is to fail under different conditions,' said Andrew. 'In an extreme scenario we expect a mean of 13 transformers to fail out of a total of 1500 transformers across the UK.' That's a cost of 13 times £5 million for the National Grid, but it doesn't believe that the actual effect on people will be very big. The National Grid has quite a lot of back-up transformers sitting at almost every node to ensure that if there are failures, new ones can be installed as quickly as possible. There is a lot of redundancy built into the system that goes back to the nationalised days of the Central Electricity Board. Where most networks might have one or two spare transformers, the early engineers of the UK grid, just to be safe, decided to have six or seven. 'We don't expect to lose so many transformers that we can't provide power once the storm is over,' said Andrew.

This, however, is where isolated communities such as in the far north of Scotland where I was visiting may suffer more, simply because there is less redundancy in the systems up there. The communities served are small and remote and the back-up transformers are located in regions of higher population density. If a transformer were to fail in the far north of Scotland, a new back-up transformer would have to be moved into place from elsewhere in the country. The quickest this has ever been done was in 16 weeks, but most of that time was spent trying to get permission to drive the transformer along the roads to get it there. Under desperate circumstances it is likely that permission would be granted immediately to drive the transformer, but it would still take a couple of weeks to get it into place.

If a huge solar storm were to happen and cause a Scottish or even UK-wide blackout, people off-grid like the Watts would be the only ones with power. They would feel no space weather effects because, compared with the transmission grid, every wire they have in their house is high resistance and so not an attractive path for the currents from the ground to track through. This gives them an additional robustness. They may suffer small, regular outages but they can ride out the big problems. It has already happened for reasons besides space weather. A recent period of bad storms and tremendous winds caused a grid blackout. Everybody in Caithness was out of power for three to four days apart from the Watts.

Fortunately, day-to-day auroral activity doesn't cause too much trouble. Even in northerly locations under the auroral oval, technological systems such as power grids have developed gradually under those 'normal' conditions so that they operate happily even under a greater strain than those at lower latitudes. It is the rare, abnormal, strong events, which create conditions much outside of the standard range, that cause the problems, and, in a sense, we are all facing extreme values together, regardless of latitude. The study of space weather is largely concerned with forecasting and warning for these extreme conditions. The more we rely on technology, and the more susceptible our technologies are to space weather effects, the more important the forecasting of these random, chaotic and turbulent events becomes.

* * *

WHEN I LEFT Lossiemouth the following day I met another ex-military pilot for coffee before I flew home. Arthur Milnes, the Chairman of the Highlands Astronomical Society, told me a story about the first time he ever saw the aurora.

'I was an advanced student pilot,' Arthur began. 'We were finishing our training on an operational aircraft and we had to master the art of night formation flying. This is something else!' he chuckled. 'Getting feet away from another aircraft in the dark is something that has to be practised.

'Three aircraft took off, with the instructor in the leading aircraft. We were then told to approach the other aircraft very cautiously from astern, closing the distance carefully, moving in slowly. Well, it proved incredibly difficult.

'At night you are usually flying on instruments, but we were also looking out at the other aircraft, at the tail lights we could see ahead. So we were shifting our gaze in and out of the cockpit.

'It proved so difficult that the instructor finally said, "Stop, stop! Move out, we're going home!" He had realised that there was an aurora that was causing false horizons. Usually the horizon is defined by a faint glow of some sort and, as a pilot, you pick up information from that, almost subliminally I think. During that formation flying exercise, when we looked out we would automatically level with the horizon, but it wasn't correct. We were having a lot of trouble caused by a quickly changing aurora seemingly altering the horizon.

'As we were flying back home we were able to look out at the display without worrying about running into the instructor. Then one could see it – flickering, faint green lights like tendrils in the sky.'

That was 1957 and Arthur's first experience of the aurora.

Arthur's talk of formation flying made me think about teamwork and how, in such a dangerous situation, it was imperative to have implicit trust in the discipline and skill of your wingmen, your team-mates. It made me realise the importance of trust, particularly in a survival situation.

I thought about Svalbard and tried to imagine what it would be like skiing out through the frozen emptiness in a twilight semi-darkness, guns and flares strapped to our pulks in case of encounters with polar bears. Would we see any? Would we scare them off? How cold would it be there in February? Would we be warm enough? Would we have enough food and fuel? Could we navigate correctly? Would we be okay? I certainly hoped so, but I was glad I was going into that situation with someone I knew and, importantly, someone I trusted.

* * *

SOLAR STORMS FOLLOW a basic pattern, so although all storms are different, we have a good idea of what to look out for and what to expect when one begins.

Initially a large, complex sunspot group develops on the Sun. Then a solar flare can occur – a bright burst of radiation comprising radio and X-ray wavelengths as well as intense visible light. Highly energetic particles are also released with speeds sometimes approaching that of light. Often a coronal mass ejection is also triggered, which releases hot plasma into space and may or may not hit the Earth, depending on its direction.

Not all sunspot pairs are likely to produce flares. Forecasters look for complexity. Joe Allen, an eminent engineer who spent his career at the NOAA Geophysical Data Centre in Boulder, Colorado, told me a story about predicting when sunspots would blow. Joe was one of the first people to realise that satellite problems were associated with solar and auroral activity, and was working in the field of space weather since before it even had that name (from about mid-1972). He was, in fact, one of the first people to use the term 'space weather'.

'I remember a young Air Force officer in the forecast service who had a reputation among his fellows for being a top predictor of which spots or spot-groups would produce big flares,' Joe told me. 'He took so much kidding about his skill that finally he confessed that he looked for "cat's whiskers" like Felix the Cat.' When he noticed more spot complexity, indicative of more tightly twisted solar field lines emerging at the surface, he predicted that the region would be more likely to flare. The officer built a good reputation until everyone began looking for the same appearance. However, it is still not possible to predict when a solar event like a flare or coronal mass ejection will occur, only to say when one is more likely.

Solar flares are a burst of light. The X-ray, UV and radio wavelength elements can affect navigation and communications, but by the time we see the flare the effects will already be with us. Monitoring the Sun is critical so that we can assess the likelihood of a flare. For the effects caused by highly energetic particles we may get around ten minutes' warning. There is not much time to react and so, like for flares, providing alerts is not really possible. With a coronal mass ejection we have a little longer. Depending on the speed we could have a day or three's notice that something may hit us, then about a half hour warning when we know exactly whether or not it will and if the magnetic field direction is right to cause a big storm. Here, careful monitoring and forecasting are of great importance for power grids and satellite operators so that they can take precautions.

When a coronal mass ejection happens, a vast blob of plasma is launched from the Sun into space at speeds of hundreds of kilometres per second. Whether it hits the Earth or not depends on where and when it was launched, its direction and its speed. Whether it will couple with our magnetosphere and be geo-effective depends on the direction of the plasma's magnetic field (southwards will cause storms). Predicting whether it will hit is not straightforward because we can't know enough of the details we need for modelling until the plasma gets closer to Earth and we can measure its properties.

The ejected plasma does not come straight to Earth, even if it is launched in our direction. The Sun is rotating, so everything coming off it (ordinary solar wind as well as coronal mass ejections) gets twisted into a spiral shape as it leaves, just like when a girl spins with a long ribbon and the ribbon spirals out to encircle her. The ribbon is like a blob of plasma over time. Trying to figure out if this spinning plasma blob will hit us, using only a picture of the plasma being ejected, is understandably tricky. Fortunately, people do try.

Affectionately known as Swipsie to employees, the Space Weather Prediction Center (SWPC) is arguably the world centre for space weather. The Forecast Office provides 24/7 monitoring of the Sun and the solar-terrestrial environment, and alerts important registered users such as NASA, power grids and satellite operators to dangerous changes in space weather conditions. SWPC is a descendant of the Central Radio Propagation Laboratory, set up after the Second World War for ionospheric monitoring to improve radio transmission, and now comes under the NOAA Weather Service. It is located in Boulder, Colorado, but it provides information globally and works with other national space weather agencies, such as the new UK Met Office Space Weather Operations Centre (MOSWOC).

The Space Weather Prediction Center assimilates multiple sources of satellite and ground-based data which it uses as inputs to models to make predictions of geomagnetic storms. Primarily scientific satellites like ACE* and SOHO† work alongside operational

---

* NASA's Advanced Composition Explorer satellite.

† ESA and NASA's Solar and Heliospheric Observer satellite.

weather satellites like the geostationary GOES*. SOHO is a satellite from 1995 that studies the Sun from core to corona. It has a coronagraph instrument that uses a black disc to block the bright Sun and so study the faint corona and solar wind plasma, including taking pictures of coronal mass ejections. ACE was launched in 1997 and sits close to the Earth, directly between us and the Sun, at a stable point called L1. It measures properties of the solar wind plasma as it passes, and this is the first time that we learn important information about the plasma about to hit us, such as the magnetic field direction, its speed and its density. As well as monitoring terrestrial weather, GOES has a solar X-ray imager to watch for flares. Other satellites are also used whenever possible. In recent years the two STEREO satellites have been important for side views, but they can't always be relied upon because they cross behind the Sun out of communication from Earth.

On the ground, SWPC makes use of the global network of magnetometers measuring the variation in the Earth's magnetic field, receivers that measure ionospheric disruption via scintillation and neutron monitors that give an indication of incoming fast-particle radiation.

Using all this data and more, SWPC provides 24/7 monitoring of solar conditions 365 days a year from its forecast office. With two forecasters on duty at any one time, they monitor the current status. Alarms go off if a solar flare is detected or if, for example, solar wind speed exceeds a certain threshold. If anything serious happens, the forecasters review the solar, plasma and magnetic data and generate alerts where appropriate. Priority customers with critical operations are on the phone-notification list, such as the power girds, the Air Force and NASA. Others get email alerts, and, for those in remote areas who can't get email, SWPC sends out space weather broadcasts on HF radio every three hours. They only appreciated the importance of this service recently when they tried

* The NOAA Geostationary Operational Environmental Satellite.

to discontinue it, assuming everyone was now on email. News of the proposed cut was met with an outcry.

Alerts go out about X-ray and radio bursts, energetic particles and geomagnetic storms. The alerts for geomagnetic storms come in three stages: Watch; Warning; and Alert. 'Watch' is an initial heads-up with a lead time of fifteen to seventy-two hours depending on how fast the ejected plasma is moving. It is issued when the SOHO satellite sees a coronal mass ejection and computer modelling (using its estimated speed and direction to track its path outwards on the solar spiral) shows it could hit Earth. A 'Warning' is issued if the coronal mass ejection does indeed reach Earth and is detected by the ACE spacecraft to have a southward magnetic field, meaning that it will cause a geomagnetic storm. This gives affected industries up to an hour to prepare. The 'Alert' is a real-time indication that the storm is now upon us.

Providing this kind of round-the-clock coverage is no mean feat. Just to be able to receive continuous uninterrupted data from one satellite requires a network of receiver stations around the globe – for ACE these are in Japan, Korea, Germany and America, and there is a network of back-up stations too. It requires global cooperation.

SWPC provides its data online to all who care to view it. As well as the raw data and its forecasts, it rates the incoming conditions on various scales. One of the best used of these scales is the K-index, which is often displayed on aurora enthusiast sites as an indication of the strength of aurora to expect. It is a planetary average of magnetic disturbance on a scale of 1 to 9 and is used to characterise the strength of magnetic storms. The initial, K, comes from the German word '*Kennziffer*' meaning 'characteristic digit'. The index was developed by the German geophysicist and statistician Julian Bartels in the 1930s and has been used by SWPC since forecast operations began.

As solar wind plasma hits the Earth's magnetosphere and

disrupts our magnetic field, magnetometers across the world detect the disturbance. Every three hours, each magnetometer station assesses how much the magnetic field signal fluctuates, obtaining a measure of swing between the smallest value and the largest value recorded. This swing is then assigned a K-index value – 1 being very little swing and going up in bands to 9 being large fluctuations. This is a local K for the specific magnetometer station.

The planetary K-index, or Kp, is calculated as a weighted average of K values from thirteen mid-latitude stations and gives and indication of the global disturbance. SWPC provides an *estimated* value of Kp on its website. The estimated value is necessary simply because not all the magnetometer stations have real-time data links to SWPC. In order to provide a real-time 24/7 indication of storm strength to customers, a subset of magnetometers must be used, which is what gives the estimated Kp displayed online.

SWPC then goes one step further to produce the NOAA G-scale, which shows the significance of geomagnetic effects to industries likely to be impacted. The G-scale is a one-to-one correlation between the top-level bands of the K-index: Kp of 5 corresponds to G1; Kp of 6 to G2; up to G5 at the Kp of 9. A Kp value of below 5 is insignificant and not classified as a storm. People only really start taking notice when we hit the G-scale.

For users who want more detailed information, observation data is openly available and SWPC provides access to the models used for forecasting, such as WSA-Enlil or the OVATION Aurora Forecast model. WSA-Enlil uses plasma physics equations and understanding to predict the solar wind conditions as it moves through the heliosphere. Its output is pretty, coloured spirals indicating the density and speed of plasma emanating from the Sun. The OVATION Aurora Forecast model gives a prediction of the auroral oval showing its position above the globe and estimated intensity of the aurora.

There are other models of fast particle impacts or ionospheric disturbance interesting to those industries that will be affected.

The thing to remember with all this data is that it is just predictions. Models are not perfect and they rely on having accurate inputs – and that depends on the data we can get from satellites, and when we can get it.

Space weather has its roots in astronomy but there is increasingly an input from meteorological research. Many techniques from terrestrial weather forecasting are crossing over and proving very useful in space weather.

One such technique is called *ensemble modelling*. Forecasters run a model many times in parallel, varying the input parameters (the starting conditions) only slightly to see if any small changes in the inputs could make a dramatic change in the forecast. This gives them a probabilistic feeling for what the risks are. They make these small variations because all the inputs to the model are based on real measurements of, say, surface temperature or water vapour or air pressure. There are uncertainties on all these measurements, meaning that they could actually be one of a range of values. Running multiple simulations within the range allows forecasters to explore the implications of the uncertainty and see what could happen if the measurement were actually just a tiny bit different. In chaotic systems, small changes can have big effects.

An important driver for this approach in the UK was the weather forecasters missing a big storm that struck in October 1987. I remember it because a lot of trees fell down in the woodland near my childhood home, which made great dens for a while until they were all cleared away. It also caused power cuts, traffic chaos and eighteen deaths. But no one saw it coming. It was just outside the forecasters' predictions. In those days they made only three or four computer runs with different input measurements to see what might happen; now forecasters make dozens of runs daily. If they had used ensemble modelling in 1987 they would have known that there was a risk of the big storm and they could have lessened the impact by warning people sooner.

Another method used heavily in weather forecasting is data

assimilation. This is where forecasters continually feed up-to-date measurements into the model to keep it tracking the real world. It gives an adjusted forecast that fits to current observations and allows longer-range forecasts to be modified and improved in accuracy as an event approaches.

Space weather forecasters are now trying to apply these meteorology techniques to space weather to assign a probability to the risk of a really big event. They seek improved ability to predict where fast plasma from a coronal mass ejection is travelling, what the likelihood is of it hitting the Earth and, importantly, what is the likely orientation of the magnetic field (is it southwards?). It is also useful to know if the coronal mass ejection will interact with the solar wind ahead of it because this can twist up the magnetic fields and lead to a more severe event on Earth. Many of the largest geomagnetic storms on Earth have been the result of several coronal mass ejections arriving one after the other.

Yet it is all very well saying what space weather forecasters want to achieve, but the trouble is, we need more data to do this. Fortunately, there are plans to improve the situation. Space weather is gradually moving from a scientific basis to a more operational basis, like the weather service. Rather than scientifically motivated missions collecting data that may also prove useful for space weather, they need dedicated satellites supplying real-time data for the models.

Bringing in true operational capability for space weather will be a global effort and, again, one that can learn from meteorology. I spoke to Terry Onsager, Director of the International Space Environment Service (ISES) at SWPC, about the issue.

'Space weather affects the entire globe, but the impacts can differ widely from place to place,' Terry told me. High latitudes suffer greater geomagnetic activity while the equatorial regions may experience greater ionospheric disturbances. This means that we need detailed measurements from around the globe of conditions like the state of the ionosphere or the local magnetic field and how these are changing, as well as measurements of the Sun so that we know what is coming at us.

Only a few countries need to participate in the space-based measurements, but this is expensive so the more they can share the load the better. 'Currently there is not a good way to do this,' said Terry, 'but it is a good goal. The challenge is to get countries around the world contributing local measurements, and to get enough of the large, space-capable countries participating to share the cost of the space-based observations.'

With regard to forecasting, increasing numbers of nations around the world are engaged in space weather research and are developing services. Collaboration between these countries will rapidly improve global capabilities and understanding.

In this vital and global issue we can look to our experience in terrestrial weather forecasting, which developed in a similar way. The World Meteorological Organisation provides the connectivity for the observing systems for terrestrial weather around the world. Using this network, with an agreement amongst the members to use common formats and common metadata, data is accessible and available wherever it is taken.

The space weather community is working towards having that same capability through the World Meteorological Organisation by including space weather observations into the WMO global observing system, where the communication infrastructure is already in place. The WMO has a plan for the years between 2016 and 2019 to integrate space weather efforts into their core programme.

'In time we will have a good, coordinated system to access space weather data from all around the world,' said Terry. 'Right now, one of the most important things we can do is encourage as much engagement as possible between the space weather service providers around the world. Being aware of our capabilities and common needs will allow us to improve together.'

At the same time as we work to improve global coordination, we also need to be planning future space missions to ensure we have the data we need when we need it. Currently, most of the important satellites for space weather forecasting are scientific research missions. Satellites like SOHO and ACE are critical, but they are one-offs for science and have no back-ups, so losing them would mean losing huge capability. Operational missions – as opposed to scientific research – also need more robustness and resilience in the system, with back-up ground stations, extra antennae and spares on the satellites to minimise failures.

The United States' first operational space weather satellite in deep space was launched in February 2015. The Deep Space Climate Observatory, or DSCOVR, replaces ACE as the SWPC real-time data provider from just outside the Earth's magnetosphere. DSCOVR is a partnership between NOAA, NASA and the US Air Force. Its primary role is as a NOAA space weather satellite but it also carries some NASA instruments for Earth and space science.

Like ACE, DSCOVR orbits a Lagrange point called L1, which is a point in space where the gravity between the Sun and the Earth is perfectly balanced, so a craft gets pulled neither one way nor the other. L1 is about 1.6 million kilometres (1 million miles) from Earth on the line between us and the Sun, so anything that hits a spacecraft at L1 will also hit Earth between fifteen minutes and an hour later depending on its speed. SOHO also sits at L1, taking pictures of the Sun without ever being blocked by the Earth. Plasma and image data from L1 is a vital part of the SWPC forecasting capability.

However, in order to get the increased prediction performance that forecasters desire – such as better accuracy in coronal mass ejection speed and arrival times, and earlier warning of the possibility of solar flares, fast particles or fast plasma from coronal holes – we need an additional, different view of the Sun from space. Plans are being put forward for a future operational mission to another Lagrange point, L5.

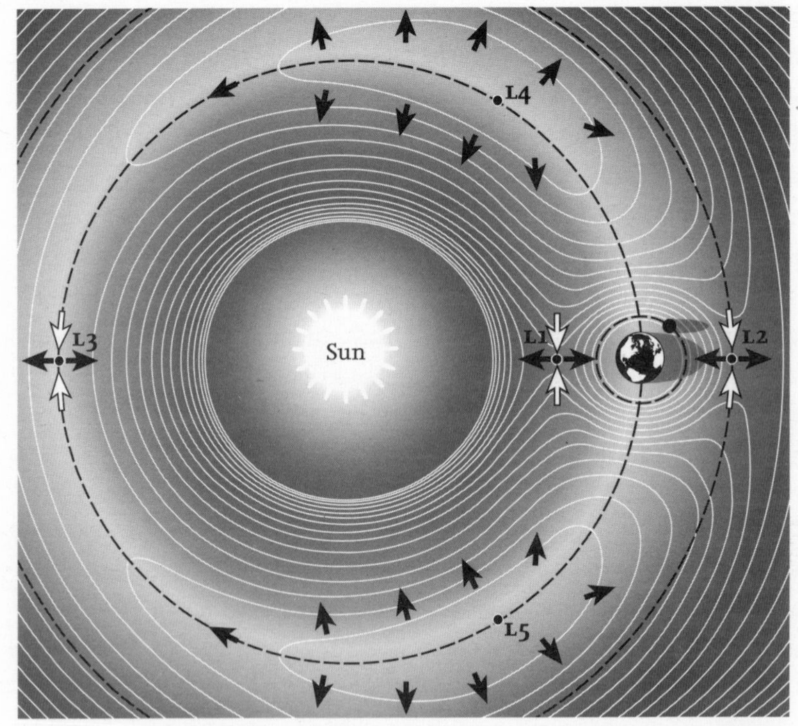

Diagram of the gravitational wells of the Sun and the Earth and the resulting Lagrange points where spacecraft can remain stable. ACE, DSCOVR and SOHO orbit L1, and a new mission is being planned to L5.

L5 is out to the side, away from the Sun–Earth line, orbiting behind the Earth. It's as if the L5 point is the third vertex of an equilateral triangle drawn with the Sun and the Earth at the other corners. There is another similar point, L4, ahead of the Earth. You can imagine these gravitational balance points by thinking about the Sun and the Earth as balls lying on a flexible rubber mat. The large Sun and the small Earth will lie in the bottom of wells made by their weight. Between the Sun and the Earth will be a saddle point – L1 – and out to the sides will be two hillocks where the edges of the two wells push against each other – L4 and L5.

A view from L5 would bring numerous advantages, the foremost of which would be a better estimation of the direction and speed of coronal mass ejections as they are launched from the

Sun. Seeing a large ejected plasma blob when looking directly along the Sun–Earth line is like seeing an object in silhouette. It can be difficult to tell if the object is a small thing nearby or a large one far away. Similarly, a slower, wider plasma blob looks the same as a faster, narrower one.

Having a side view gives us a measure of depth. Tom Berger, solar physicist and Director of SWPC, compares it to having a second umpire at a cricket game. The first umpire stands behind the bowler looking down the pitch and can judge leg-before-wicket and wides; the second umpire stands to the right, about level with the batsman, judging run-outs and stumpings. Without that side view it would be hard to assess fairly.

Tom went on to tell me about models that have been run that show the coronal mass ejection just glancing the Earth, but which have then surprised us by arriving earlier than expected and actually hitting us. We can't make sufficiently accurate predictions without a good measure of the speed and angular extent of the ejection. The SWPC forecasters make several model runs considering different scenarios that match with the chronograph picture, but they can give very different results. The forecasters must make a decision and issue their prediction, but they only get enough data to make a real warning when the plasma hits L1, giving us an hour or less of notice.

As well as more accurate speed and direction information, a satellite at L5 would give us about three days' advance warning of what is spiralling to Earth because it would hit that spacecraft first. It can also see further round the side of the Sun because it is sitting out to the side, so would give us a view of the sunspots and coronal holes rotating round to face us almost four and a half days before we see them from Earth. With this vantage point we could be better prepared for what might come our way.

The NASA STEREO spacecraft have been used to provide side views before, but this pair of scientific satellites has been looping around the Sun so they are not always in range or in the right place to provide data. However, they proved the worth of a different perspective. Now an international consortium led by the US SWPC and the UK MOSWOC* is planning an L5 mission focused on forecasting.

'L5 was originally planned solely for science,' Tom told me, 'but now it is being discussed internationally as an operational mission because it has been realised that we really need this for forecasting. We also want to get some science out of it too, so the data will be made available for science as well as used for operation.' The planning meetings actively involve the scientists so that the entire space weather community can get the most out of the mission as possible. But there always has to be a balance between cost and weight and ambition.

'Ultimately,' said Tom, 'there is enough high-level support for operational missions that we don't necessarily need the scientists' approval, but if we can collaborate with the scientists to enhance the general understanding and make the prediction and forecasting better in the future then that is worthwhile.'

* * *

WHILST IT IS clear that having good forecasts is essential for combating the detrimental effects of space weather, knowing how to use the forecasts is equally important. As Professor Sir Mark Walport, Chief Scientific Adviser to the UK Government, said to the space weather community, 'I urge you to work out not only how to tell people something is happening but also what to do about it.' Users and industries affected need to work with others to determine how they will respond.

But what can be done? How can vulnerable parties like the National Grid or satellite operators protect their assets in the face of a solar storm?

---

* MOSWOC – the Met Office Space Operations Centre.

Andrew Richards from the UK National Grid told me that they work closely with the MOSWOC, which works closely with SWPC but can give UK-centric advice. Andrew receives the Met Office's daily space weather forecasts and collaborates with them on how to improve services.

It takes a big storm for the National Grid in the UK to be disrupted, but that doesn't mean it doesn't have to take precautions. Storms need to be Kp9 (G5 on the NOAA scale) before the UK Grid feels any effect. At this level, the aurora may be visible across the entire country – or maybe only half the country, or maybe even into France . . . The trouble is that the Kp9 band does not have an upper boundary, so saying it is a Kp9 or G5 storm indicates that it is big, but gives no sign as to just how big. That's not much use for the National Grid.

In the UK there have been Kp9 storms where nothing has happened, so the Grid really needs Kp10, 11, 12 measurements. In fact, they categorise disturbances according to their own 1–5 scale within Kp9. Category 3 or 4 is when things start getting interesting, with a National Grid Category 4 storm a bit like the 1989 event in Québec. The National Grid also needs to know about fluctuations on a shorter timescale than the three-hour window used to calculate the K-index. They already use measurements taken every fifteen minutes by the British Geological Survey for their own calculations of the magnetic disturbance. It may be that the K-index, historically introduced by scientists studying early space weather effects when nationwide electricity grids were not well established, may require modernisation to fit the needs of the end user.

If a coronal mass ejection does hit Earth, one thing that can be done to reduce the chance of damage to the National Grid is to stabilise the system against voltage swings. This means getting as many power lines in operation as possible so that the network is highly connected and so that stray induced currents can disperse. The Grid has 8000 kilometres (5000 miles) of power lines around the UK, but not all will be working all the time. There are always lines that have to be repaired or are undergoing general maintenance work. If a large solar storm hits, the National Grid will want to get all work suspended and get the lines back into use. The problem is that getting all the lines back requires about four days' notice.

If a coronal mass ejection happens on the Sun then we don't have four days' notice of impact – we have perhaps one or two days. It is not enough to just wait until a solar event occurs – the Grid needs to start planning five days ahead and taking precautions just in case something happens. They know that the very large coronal mass ejections are associated with large flares which, in turn, are associated with large sunspots – and sunspots take time to rotate around the Sun. So they monitor the Sun and keep a lookout for large, unpleasant-looking sunspots on the left side that may be liable to flare. If they see one, they must start taking action because it will only be about four days before the sunspot is in the danger zone in the middle of the solar disc (the part of the Sun facing Earth). If a coronal mass ejection were to occur there it could cause trouble on Earth.

'I estimate that 90 per cent of the time we take action it will have been to no avail,' said Andrew. 'It's an unpleasant cost to carry, but the consequences to our reputation and to our bank balance is such that we think it is worth doing.'

Additionally, if a solar storm hits, rather than turning things off, the Grid turns all of its high-voltage equipment on. With more equipment switched on, the burden of the excess current flowing through the system can be shared so, hopefully, nothing is overwhelmed and damaged.

The National Grid would really like to have better predictions of whether or not a coronal mass ejection is likely to happen. After that, they would like a prediction of the size of the event, which depends on the density and speed of the plasma (how much stuff is coming and how hard will it hit us), and a prediction of whether the magnetic field will be southwards and therefore able to open up the shield of the magnetosphere. It would also be useful to be able to model how the magnetic field will change at ground level based on what is hitting the Earth.

These are all items on the National Grid's 'wish list'. Certainly a mission to L5 will help with some of them, particularly in getting the data on the plasma conditions. For the others, missions, observations and experiments building the core science understanding of solar, plasma and magnetospheric physics will put us in a position to provide better predictions in the future.

The situation for the satellite operators is different. The satellite environment is hostile and damaging, so everyday normal space weather is important as well as big storms.

'Energetic particles are like bleach or corrosion for a spacecraft; they cause progressive weakening of certain systems over time,' Ewan Haggarty of the Spacecraft Management Authority of the Airbus DS SKYNET programme told me. 'Good practices and mitigation strategies are already in place for day-to-day contingency situations, so dealing with space weather is an extension of what our Spacecraft Controllers already do.'

Unfortunately the forecasting of extreme solar particle events, which can damage satellites, is in its infancy. A solar flare sometimes precedes energetic proton arrival, but not always. In any case, the observation of a flare might only give thirty minutes' advance warning time, or less. Detecting relativistic electrons at the L1 point (between the Sun and Earth) might also give an indication that a host of fast protons is on the way, but for protons that are also travelling close to the speed of light the warning time would be only about eight minutes. This does not give much time for action.

Satellites do have systems to mitigate the effects of solar particle events. Multiple levels of redundancy are built into their vital systems, such as two, or even three, circuits doing the same thing. Errors in on-board computer memory can be automatically detected and corrected.

Ultimately, spacecraft have back-up modes designed to ensure their survivability. They will shut systems down automatically or enter back-up operational modes if there are significant problems with, say, power, orientation or temperature. Some satellites also have particle detectors to monitor their space environment and will take programmed action if particle fluxes get too high.

Ewan told me how at certain times, such as when an extreme storm is on its way, a human controller could take pre-emptive action to protect the spacecraft. For example, they could put the satellite systems into a simpler or more stable state or, if the energetic particle situation became extreme, perhaps even switch systems off to protect them. This would disrupt the satellite's service and impact the customer, which is something all operators strive to avoid, but protecting the spacecraft is always the first priority. Controllers take action to maintain the life of these expensive – and now almost indispensable – assets, even if that means customers experiencing an interruption.

As experience in building space-worthy systems has grown, the materials used in the electronic components of satellites have become more refined. Spacecraft now use insulating materials that do not build up an internal charge when bombarded by energetic particles. This prevents the short-circuit discharges that can result in damage to electronics or spurious control signals. Satellites also have protection against electric discharges on their surface. Satellite structures are carefully bonded electrically to prevent large voltage differences between the component parts, which would result in damaging currents flowing.

Satellite protection is especially important for the operational space weather satellites which will sit in the stormy plasma flow for their entire lifetime. As Tom Berger, Director of the Space Weather Prediction Center, says, 'you don't want your space weather satellite being taken out by a space weather event.' The extra levels of protection required for this assurance increase the price tags of already costly space missions.

Space weather can also pose a different problem for scientific research satellites, though one that doesn't interfere with the lifetime of the spacecraft. Space weather conditions can affect the taking and reliability of data, even when what is being studied is well outside the solar system.

One of my good friends, Jenny Carter, a physicist at the University of Leicester, suffered this problem during her doctorate. She was using data from the European Space Agency's XMM-Newton X-ray telescope to study an object far away, outside of our galaxy. When the scheduled observing time came, the telescope moved to point at the distant object and got ready to take data, but the particle detectors registered stormy space weather and automated systems stopped operations in order to protect the telescope. It was only on the third try, about two days later, that the telescope was actually able to take data. It turned out that a coronal mass ejection had caught up with a fast solar wind stream and was playing havoc with the near-Earth space environment.

This can be a problem for astronomical researchers who are only given short periods of observing time. Schedules for satellite telescopes are made months in advance and planned to point at certain areas of the sky at certain times of the year, and in an order that avoids swinging the telescope around too much. With such short windows of opportunity, some unlucky researchers can end up with no, or unusable, data if space weather conditions are bad. Estimates suggest that overall these scientists lose about 30 per cent of their data this way. Even though satellites give astronomers the facility to see out into the distance without the obstruction of clouds or the Earth's atmosphere, they still do not have a completely uninterrupted view.

For the most part, satellite-affecting space weather problems are an issue for the operators and don't feed down to the public level, but there is a lot of work going on 'behind the scenes' to ensure this is the case.

* * *

THE PUBLIC'S BIGGEST interaction with space weather is via the aurora. The products and alerts provided by SWPC can be equally useful for space weather enthusiasts and aurora tourists, besides informing multi-billion-dollar industries of potential risks. Various local organisations around the world send out aurora alerts by email or social media to give their followers the best chance to see the aurora wherever they are. They make it easier for people to experience that all-important connection that James Pugsley was talking about in Yellowknife.

As well as AuroraMAX, James started the Northern Lighthouse Project in Yellowknife, which aims to raise local awareness of geomagnetic storms. Small model lighthouses around town flash blue, green or red depending on the prevailing space weather conditions. James hopes he can help the community, and particularly young people, to understand geomagnetism and some of the basic processes behind the aurora. The availability of space weather data through SWPC is crucial to James' ability to do this.

James loves the WSA-Enlil model with its spiralling rainbows of plasma. He thinks it is an incredibly valuable tool. Based on the WSA-Enlil predictions, AuroraMAX sends out alerts if there will either be a direct hit or a glancing blow. But the predictions can sometimes be wrong, and James doesn't want to get people's hopes up too early.

For James, the alerts for a glancing blow are the trickiest to write. The wording has to be vague to notify readers of a possibility that may never ensue.

'Something like, "There is a possibility of a potential impact that could be happening sometime within the next 24 hours, we think, maybe".' James joked with me when I was in Canada with him.

When aurora spotters don't understand the origins of the aurora or its inherent unpredictability they can suffer great disappointment if they don't see anything.

'You should see the emails I get,' said James. 'They say things like, "You don't know what you're doing! I stayed up all night." I try to help them understand.'

It's not always bad news, however. 'I also get really happy emails when it works out,' grinned James. 'We give people the opportunity to see something amazing.'

For James, just as for the power grids, an earlier, more accurate prediction of what is incoming would be a boon. James would be elated if WSA-Enlil could be updated more frequently and the prediction improved. A new mission to L5 would enable this, helping the aurora spotters too.

In Caithness, a Kp of around 5 will give a nice overhead aurora, according to Gordon Mackie, but with a Kp of 3 the lights could be seen high on the northern horizon. The Caithness Astronomy Group regularly check the aurora forecasts and will communicate between themselves, informing each other of the clear weather spots around the county. If there is a good display in the offing there can be a lot of local interest, especially in more recent years now that social media helps the news to spread quickly.

That evening at Castlehill Heritage Centre, Gordon had showed a picture of the aurora where there were numerous people out enjoying the spectacle. It was from February 2014 when there was a big display and – as Gordon put it – most of Caithness was out. Even at one o'clock in the morning people were coming and going. He even saw a lady in her car, watching the aurora from inside. At first he wondered why she would not get out of the car to get a better view, then he realised that she had come out in her pyjamas and dressing gown.

'There were more cars there, on a cold winter's night, than I had ever seen there even on a clear summer's day,' Gordon told me.

That is the power of the aurora, combined with a forecast of high activity and the stirring of social media to increase the excitement.

On my second night in Caithness, after my afternoon with the

Watts and my walk to the Duncansby Stacks, I went out with the astronomy group to see the aurora. The activity that night wasn't high, the K-index was about 2 only, but the aurora was there on the northern horizon. A number of us stood out by the water behind the Castlehill Heritage Centre and looked out. The aurora was faint and pale, looking more like a sea fog than a heavenly dance. I could just about make out feeble pillars of light rising into the air. With the naked eye I could not see any colour, but my camera could see bands of green and red and pink, and millions of incredible stars. It was easily the best photograph I had ever taken of the aurora. That I was out with a group of auroral photographers probably had something to do with it, yet I attributed this success also to the fact that I was not freezing, merely a little chilly. My patience with camera equipment in sub-zero temperatures is thin.

Chris took a photograph of me looking out across the water at the aurora.

'Do you think of yourselves as aurora hunters or aurora chasers?' I asked between photos.

'Nah,' said Gordon, 'we're just enthusiastic about seeing it.'

'Aurora hunters sounds cool though,' Chris mused.

'But we are not actually aurora hunting. We know it will be there and we know it will be north. Clear sky hunters more like!' laughed Gordon.

Maciej, who has been featured in the press a few times, said, 'Yes, I have been called an aurora hunter, but I am just an enthusiast. If I would chase the aurora around the globe then maybe I would be an aurora hunter; if I was travelling every few months to Iceland or Alaska or somewhere. Most of the time I am just here and it happens, so I will go out and shoot it.'

'We watch it from our back door up here,' said Gordon. 'You are more an aurora hunter than we are.'

CHAPTER EIGHT

SVALBARD – SEEING THE LIGHT

DREAMS ARE JUST playthings here, where the wind strips and tears, and a frozen heart has the energy only to beat, not to feel. We survive rather than live. It could have been a beautiful story, but up towards the end of the world imagination can be far from reality.

I was in Longyearbyen on Svalbard, the Norwegian archipelago that sits in the Arctic Ocean east of Greenland between 74° and 81° north. I had come to experience the northern lights in the remote wilderness, to get a flavour of what it must have been like for Arctic explorers over a hundred years ago making forays into this inhospitable land. 'Svalbard today looks almost like a landscape from the end of the Ice Age,' says the Svalbard Museum. There is less than ten per cent vegetation, and more than half its land-mass is covered in glaciers. There is permafrost in the far north where only the topsoil thaws out in summer. There are long Arctic nights – darkness from October to February – and fjords freeze over. It is a land of ice and rock and wind.

Svalbard was discovered by the Dutch in 1596, a time when young seafaring nations competing for trade began exploring the north. Two vessels sailed from the Netherlands that year, captained by Barentsz and Rijp, and they came upon the uninhabited islands in June, believing at first that the archipelago was part of Greenland. Declaring that it was 'nothing more than mountains and pointed peaks', they named the land Spitsbergen, meaning 'pointed mountains'. By this name the islands were collectively known for the next four hundred years, until they passed into Norwegian jurisdiction in the early 1920s and were renamed Svalbard, the largest island of the group retaining the name Spitsbergen for itself alone. The new name for the archipelago, Svalbard, is an old Norse term meaning 'land of the cold coasts'.

Almost immediately after its discovery at the end of the sixteenth century, Europeans began exploiting the natural resources available on the islands. The barren, desolate and frozen lands abounded with birds, marine life, polar bears, arctic foxes and reindeer. Whaling began in 1612, to provide lamp oil, which was derived from whale blubber. The whales were harpooned by men in small boats and dragged to shore, where the blubber was cut into strips and cooked down into oil in large cast-iron pots called blubber cookers. Baleen was also collected and used; the long bristle-like plates forming the filter-feeder system in a whale's mouth were used for numerous purposes requiring strength and flexibility, such as corset stays or collar stiffeners. Whaling stations were set up on Svalbard to accommodate men and for rendering the blubber to oil. Whaling activity increased rapidly, moving out from the Svalbard coast into the Greenland Sea. By the end of the seventeenth century there were two to three hundred whaling and sealing ships out in the summer months. Whalers began boiling the blubber on the ships, reducing the need for the whaling stations on land, many of which were torn down and cleared of any usable materials. Svalbard was now used for refuge and burial grounds instead.

However, by the end of the eighteenth century the whaling was all but over, with whale stocks severely depleted and the Greenland right whale* almost extinct. Activity on Svalbard moved into a new

---

* The right whale's name even comes from whaling – it was the 'right' whale to hunt. It swam slowly, floated when killed, and had a high yield of baleen and oil.

phase of hunting and trapping. Russians had been hunting walrus there since at least the early eighteenth century, rowing in on small boats to where walrus rested in groups on the ice, piercing their thick, folded skin with long lances. It was a difficult kill but walrus were still hunted to the verge of extinction. The walrus became a protected species in 1952. The heyday of the hunting and trapping was from the end of the 1890s until the 1940s, when almost 400 people hunted mainly polar bear and arctic fox, a small fox with a lustrous, often pure white, coat and a long fluffy tail.

Come the twentieth century, coal mining had begun on Svalbard and was to become the catalyst for the development of communities. In the first twenty years there were more land claims than there was land, claims coming from people of around ten countries and from adventurers as well as specialists. Disputes and arguments ensued and this led to demands for the proper administration of Svalbard rather than the unregulated free-for-all that had endured since the sixteenth century. The Svalbard Treaty was signed in 1920, coming into effect in 1925, giving sovereignty to Norway but imposing some limitations. Signatory nations have equal rights of entrance to and residence in Svalbard, and to commercial ventures including mining, maritime and industrial activities, fishing, hunting and trapping. Taxes collected may only benefit Svalbard and not the rest of Norway. No war-like purposes are allowed.

The first year-round mining operation on Svalbard was English. The Spitsbergen Coal and Trading Company set up operations and created Advent City in 1904 – across the fjord from where Longyearbyen now stands. However, production was less than expected and their stay short-lived. Longyearbyen itself was founded as Longyear City, the operational base of the American Arctic Coal Company who started mining there in 1906. The city was named after the major shareholder of the company, John Munro Longyear. The Norwegian company Store Norske Spitsbergen Kulkompani took over in 1916 and the town was renamed Longyearbyen. 'Byen' means 'town' in Norwegian. It is now the largest settlement on Svalbard and the administrative centre. There is a smaller Russian settlement called Barentsburg, a research station at Ny-Ålesund and a mining community at Svea. Together these constitute the bulk of the population on Svalbard.

In the early days of the mining activities, the settlements that grew up were male societies where workers lived in barracks. They were not family communities. It was not until the 1970s, when the airport opened and reduced isolation, that there was a shift in society. Officials and workers were able to live with their families, and cafés and shops appeared in town. In 1993 the University Centre in Svalbard (UNIS) opened. Research and education became more active and businesses connected with tourism, trade and service opened, producing a more rounded society in Longyearbyen.

Science is important in Svalbard, and one could say that Svalbard is important to science. It occupies a position at the entrance to the Arctic Ocean with large ocean currents and rich seas. The Barents Sea is one of the most productive seas in the world. Svalbard has varied and accessible geology, where the vegetation is sparse and glaciers have scoured the landscape. The climate is favourable for such a high latitude because the west coast is affected by the warm Gulf Stream. It lies under the path of all satellites orbiting Earth around the poles, and the satellite station outside Longyearbyen tracks more than thirty satellites at once, downlinking data and sending up new commands. And its position under the polar cusp makes it an interesting place to study the aurora, indeed the only place in the world where daytime aurora may be seen – only visible during intense polar night. The natural environment of Svalbard is used as a laboratory and an observatory.

Since its discovery by William Barentsz in 1596, man has been leaving traces on Svalbard. Cultural heritage – all traces of human activity prior to 1946 – is protected by law. They are the relics of human occupation of the land. Buildings like trappers' cabins or old mines, even smaller items resembling rubbish, like an old cup or spent cartridges, are sacred. In and around Longyearbyen tall, triangular wooden towers stand frozen in a long procession across the landscape. They once formed the old tramway where large buckets of coal would be transported down from the mines to the city and the port. Now all coal transportation is by truck but the towers still stand, their silent marches showing the way to the mines, all but one of the seven in the vicinity of Longyearbyen now also abandoned.

* * *

ONE MORNING IN mid-February I followed a line of these towers with my eyes as we sped by in a taxi, following them to where they ended at a small site five minutes drive out of town in Adventdalen*. There was a large circular silo and a track up to the mine entrance a little way up the mountain. It looked like a large slide from an abandoned and frozen fun-park. We sped on, passing a handful of wooden cabins and a gridded scientific array that consisted of a lot of metal sticks standing with their multiple arms held out. We saw two large radar dishes up on the mountain ahead above another mine – Mine 7, the only mine that remains operating nearby – before my companion asked the taxi driver to stop and we descended into the ice and the wind.

We were at Bolterdalen, the third valley from Longyearbyen. I zipped my jacket up high, put up my hood, hurriedly pulled on my huge mitts and stood around awkwardly trying to help as the men manhandled the heavy sleds out of the back of the taxi. We waved the taxi off, slipped on our harnesses, clipped in the sleds and snapped on our skis. Then we set our backs to the road and began pulling.

My polar guide, Svante, and I (just the two of us) were skiing

* 'Dalen' in Norwegian means 'valley'.

out towards the east coast, where the influence of the Gulf Stream does not reach and the temperatures fall by as much as ten degrees. The trip was more intense – more brutal – than I could have imagined; more than I can really explain. I realised then that out there everything becomes about survival and nothing else matters.

The first two days were simply white. The light was flat and the sky a barely-there grey. No sun, no colour, no contrast. Low mountains rose up on either side – white with high, grey rock bands. The land was windswept ice. Snow that had fallen was quickly swept away by the wind. Some lay collected in small dips or gullies, like puddles of snow rather than water. In other parts, between the puddles, there was hard snow crust. This was the easiest to ski on because the friction on the sleds was less than in snow. The trickiest was the sheet ice where there was no grip. The toughest was soft snow uphill.

We picked our way up the valley, trying to stay mostly on crust. Numerous dog teams and lines of snowmobiles passed us going up and down the valley. There is an ice cave at the end of Bolterdalen and it proves a popular tourist destination for short dog-sledding trips. A meltwater channel in the glacier is made accessible via a small hole in the snow and a ladder. The walls of the cave were smooth, clear ice, transparent as glass, with uneven layers of pebbles and dirt that showed where one-time surfaces of the glacier had been buried and pushed down. The ceiling of ice crystals sparkled under the light of our head torches.

Gradually our route began climbing and became snow rather than crust. The effort intensified. On steep sections we had to take off the skis and haul our sleds up on foot. The pulk threatened to pull me backwards downhill, so I leaned forward hard and dug the sides of my boots into the snow.

That first night we camped just over the high point of the climb and spent a relatively pleasant night with the temperature a manageable -20°C or so. The next day we descended a fairly narrow river valley, following a single reindeer for quite some way.

Every so often it would stop and turn to look back at us, pausing awhile before trotting off again. We skied over its tracks and frozen droppings. Later the way flattened and widened, and the solitary reindeer found its herd. They ran away as we skied on. Eventually we turned a corner and emerged into the wide Reindalen valley and headed east. From then on we didn't see another person until we were returning back the same way.

Reindalen was vast and beautiful – a wide, long expanse edged by flattened mountains that looked like a giant line of piled white sugar subsiding into the valley. The surface was mostly icy crust, so pulling the sleds was relatively easy, but again there were puddles of snow to break up the monotony of crust. We were accompanied always by the loud scraping sound of skis over uneven, frosty ice. It was too loud to talk. We progressed in our own individual worlds. Every hour or so we would stop for a very quick break, putting on a down jacket immediately. I would sit on my sled and drink some water from my thermos flask, eating a few nuts or a biscuit, always swapping my hands in my mitts between each action that required the dexterity of free fingers – it was the only way to prevent the fingers becoming painful from cold. Despite my best efforts they would hurt anyway, and it was always a relief to start skiing again and for the pain in my fingers to gradually diminish.

Day three was in colour. It was clear and the sky showed a tinge of blue. When the Sun rose, although we didn't see it, it licked the top of the mountains and painted them a pinkish hue. The whole sky near the horizon turned an orangey-mauve, like watercolour blood orange. We made our way quickly down the flat, lavender-tinted valley, travelling at what felt like a fast walking, or slow jogging, pace.

It got colder. By day three I could no longer write my diary because my fingers were too cold even in the tent. My pen had frozen up the day before and I had written in pencil. Now I stopped entirely. Stopping to pitch camp in the evenings was a race to get everything set up and to get warm. As the guide pitched the tent I would get the bedding, fuel, burners, pan, food and guns ready to go in. He would put up the polar bear trip wire while I would get everything inside and dig a step in the porch for easy access, piling snow up in the other half for melting for water later. Then I'd go into the tent, pump up my sleeping mat, organise, strip off my waterproof outer gear and get into my sleeping bag to get warm and out of the way as he came in.

We lit a burner in the tent to take the edge off. It wasn't warm, but we could function. We could pass a relatively pleasant evening once we had eaten our rehydrated food and heated up water for our bottles, chatting in our sleeping bags over the small burner. Together we were relaxed and friendly, enjoying each other's company.

Mornings were unpleasant. I never enjoyed chipping away the ice from the opening of my sleeping bag and wriggling my way out into the frigid air. Svante was kind and patient. He would always emerge first and set the burner going as I summoned up the wherewithal to open my zip. It would take me about twenty minutes with the burner on to slowly emerge from my bag into the frozen air, inching out little by little to acclimatise myself to the temperature difference. Sitting up later after breakfast, with my down jacket on and my legs still in my bag, I would spend about ten minutes summoning the willpower just to change my socks ready for skiing. Eventually I would have to get out of my sleeping bag entirely, put on my chill, stiff outerwear and go out into the real cold. Yet I was never unhappy. It was just a different, sometimes challenging, experience, but not a bad one.

That third evening we saw the northern lights. We looked out at around eleven o'clock and there they were – a faint greenish white, stretched out east–west across the whole sky, reaching up in places like towers to the heavens. From where we were camped we had a wide view and it was beautiful to see the lights over the full horizon.

'Mission accomplished!' Svante exclaimed as he emerged from the tent to join me. We stood there looking up.

'It's pretty,' I said, smiling at him. 'It's moving a bit too. That bit that is now bright, by the mountain,' I pointed, 'has been moving along gradually.'

'You are lucky.'

'I'm so glad I've seen something. I thought we might tonight since it is so clear. I'm glad we stayed up.'

'Nice night for northern lights,' he said, turning.

'Yes, just so cold!!'

'Yeah, it is cold,' he said as he unzipped the tent and ducked inside.

I was still staring. 'They're nice, actually,' I mused to myself. 'Nice rays.'

'Yi-ha! Northern lights!' I could hear Svante singing to himself as he rustled his way into his sleeping bag.

I stood and watched awhile. It was right above me now and the pole star was just behind. I sighed happily, then shivered and hummed to myself as I watched a minute longer.

'It's so pretty!' I said again to the tent, 'even though it's so faint. I'd love to see more intense colours, but it's nice that it's so wide. It's like tall light towers in the sky, or rays of colour raining down. Really slow, colourful rain.'

I stamped over to my pulk at the side of the tent, the ice snapping and crunching beneath my feet as I fell through the crust into the softer snow beneath. I brushed the ice crystals off the bag with my mitt before taking it off to struggle with the zip. Scratching around inside, I found my tripod and pulled it out, hurriedly putting my mitt back on. Using the other hand, I erected the tripod, singing as I did so. With dexterity reducing I worked as quickly as I could on the adjustments of the tripod and screwing down the small camera. Ignoring the pain in my fingers, I tried to remember the settings I was supposed to use. I took a picture. Thirty seconds is a long time in pain. I looked at the image as it flashed up and I realised I still had my head torch on. I turned it off and tried again. Then the camera battery died. I quickly put away the tripod and darted back into the tent and to the warmth of my sleeping bag. I was happy.

The aurora can take many shapes and forms. There is infinite variety in the twists, the depths, the rays, the colours and the brightness. When scientists started making systematic observations it made sense to classify the aurora they were seeing into various types and subtypes for ease of comparison. Størmer published the first system of auroral classification in 1930 in the *Photographic Atlas of Auroral Forms*, which was later updated for the *International Aurora Atlas*, released in the 1960s.

Classification is by five main characteristics – condition, structure, form, brightness and colour. Form is often the most obvious. The aurora may be discrete, appearing as an *arc* or a *band*, or diffuse like a *patch* of light or a *veil* across the sky. Or it may be *rays* coming down as shafts of luminosity. The aurora's condition was defined as whether it was active, quiet, pulsating or coronal. Sub-categories defined the extent. Structure assessed the homogeneity of the form – was it smooth and regular or lined with horizontal striations or vertical rays? A common form of rayed band was often referred to as 'curtains' or 'draperies' since it assumed that hanging, wavy appearance. Brightness was an index giving an estimation of the green oxygen emission and defined by vague statements such as, 'comparable with Milky Way; green colour not perceived' for level 1, or, 'much brighter than 3; sometimes casts easily discernible shadows' for level 4. Colour was split into classes indicating whether the aurora was dominated by red and green or blue and purple.

The most familiar form is the twisted band. According to the historic classification, an arc is a simple curve of light with a smooth lower border, whereas the band is similarly continuous but with a twisted, irregular lower border. In my experience of discussions of the aurora, often the two are used interchangeably. Particularly

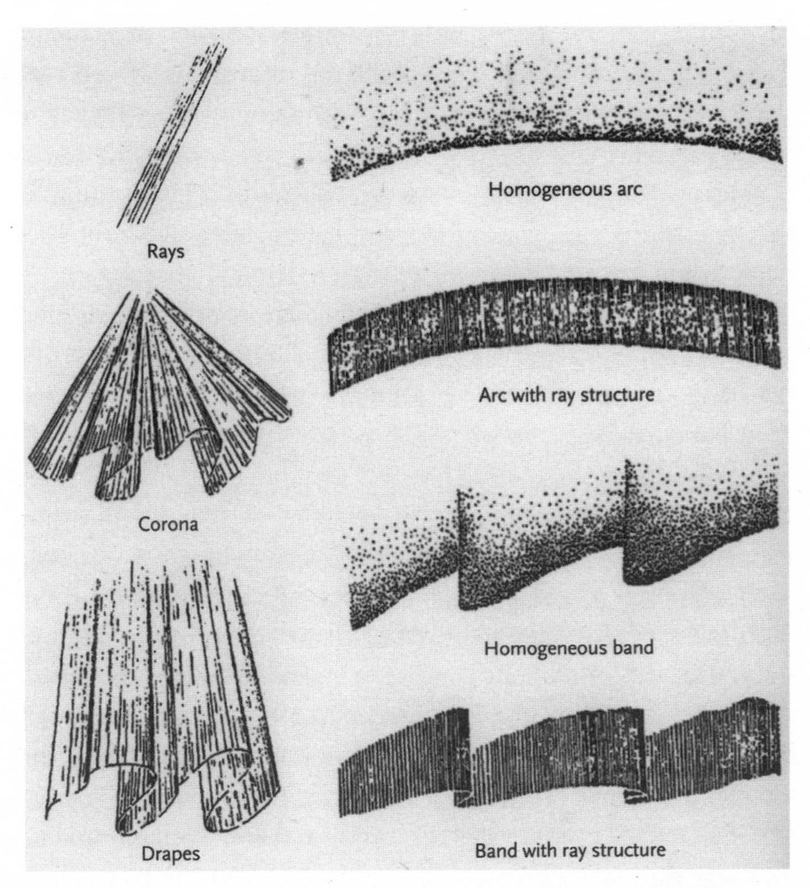

Diagram of some of the old aurora classifications showing arcs and bands. Now we recognise many of the old classifications as being similar auroral structures seen from different perspectives. Courtesy of Alv Egeland.

when discussing substorms, we talk about the arc moving southwards and breaking up.

I remembered back to my conversations with Eric in Calgary, and how he had told me that there are really only two main forms of aurora: the arc, and the patchy, pulsating aurora. These are fundamentally different physical processes, whereas several of the old-style classifications are simply different ways of looking at the same thing.

If you are lucky enough to be witnessing an auroral display, the 'type' of aurora you see depends on where the activity is happening relative to you. Mostly, what you will see is an arc or a number of arcs. A single, wide band viewed from underneath would look like an arched line in the sky. Multiple twisted, rayed arcs viewed from a distance look like curtains; but were you to look up from below, it would take the appearance of a corona, like a starburst overhead with the rays meeting above you. It's not a new type of aurora, it's perspective. Think about walking through a city of skyscrapers; as you look up, the tall buildings around you appear to converge to a zenith. Then cross the river, or take a boat ride from the quay, and look back to where you walked. Now the skyscrapers have a flattened, more regular form. The individual buildings look like rays and their different heights give the city a wavy edge. It is the same city, just a very different view.

For auroral scientists nowadays, classification of forms is not as important as it used to be. Classification according to colour (wavelength) or intensity is more valuable, because this tells us about the energy of the incoming particles and the properties of the regions from where they originate. What is important about the two major forms – the arc and the patchy pulsating aurora – is the physical mechanisms that produce them. These are still far from clear.

The morphology of auroral arcs does not yet have an explanation; what causes most of the shapes and patterns we see remains unknown. This is a big question, and it itself points to a bigger one. The auroral arc structures are not consistent with the process of reconnection in the magnetotail region alone accelerating electrons to Earth, as described in Canada. Something is missing. The energy of electrons accelerated by reconnection in the tail is around twenty times less than it should be to create the bright colours that we see. Moreover, where reconnection happens, where open field lines are becoming closed, maps onto just the poleward boundary of the auroral oval. The auroral oval itself extends over about a thousand kilometres (600 miles) in width and only one edge of that can be explained. Most of the bright, visual auroral forms that we witness simply cannot be due to the reconnection process alone. What is accelerating the electrons? There are various theories but, as yet, no agreement.

A colleague of Eric's, David Knudsen, Professor of Physics at the University of Calgary, suspects that 'there is going to be more than one mechanism causing auroral structures, but there won't be many.' Eric thinks that some of the different models put forward so far will be branches of a bigger theory.

David himself has a theoretical model that attempts to answer this question. He told me that there is some kind of 'generator' in the magnetotail which creates electric fields and currents. What is unique about his theory is the way he combines large sheets of current flowing in the same direction as the magnetic field with large-scale, convective movements of the plasma. This creates electron acceleration that would produce the structure that we see in arcs. Some of the other theories only explain a very small subset of what is observed, but David is looking for something that explains the fairly stable background process that happens to generate billions of fast electrons to create an enduring pattern over a spatial scale tens of kilometres in width and spanning longitudes (the arc).

David likened the problem to a tsunami. It is as if we have a huge tsunami wave coming towards the shoreline and all the study so far has been of the froth. No one has yet explained why this huge, massive structure is headed towards shore in the first place.

David believes that it is more than just small-scale structure added up to get the large-scale morphology. He believes there is some as yet unidentified (or unaccepted, if his theory is correct) mechanism that makes the large-scale background, which then tends to break down and become unstable, causing the motion and the dynamism that we see in the aurora.

A patchy pulsating aurora is profoundly different to an arc. The patchy pulsating aurora is diffuse, meaning it is less concentrated, and is created in an entirely different way. Rather than electrons being accelerated down the field lines, the patchy pulsating aurora is thought to occur when trapped particles are knocked out of their trapped orbits by waves travelling through the magnetospheric plasma. These trapped particles are from the region of closed magnetic field lines earthward of the tail reconnection zone. There may be some overlap with the Van Allen radiation belts – the crescent-shaped regions to the side of Earth where electrons are trapped bouncing from pole to pole – which are much closer to Earth. Knocking these particles out of their trapped orbits is called 'pitch angle scattering' and it essentially changes the position near the pole where the electron bounces back.*

You can think of an electron's pathway around and along a curved geomagnetic field line a bit like a slinky spring. It will be stretched out a bit near the equator and will steadily squash up towards the poles until it can't squash up anymore, the ends of the slinky spring finishing somewhere in the far upper atmosphere or above. The ends of the slinky are the bounce points, where the electron orbit has become so tightly squashed that the electron changes direction. The pitch angle is the angle the slinky spring wire makes with the magnetic field line at the equator.

Pitch angle scattering by plasma waves changes this arrangement. It reduces the pitch angle so that the electron at the equator is travelling more directly along the field line. This is like stretching out the whole slinky a bit. It will still compress and bounce near the poles, but now the whole spring will be longer and the bounce points will be deeper into the Earth's atmosphere. If the bounce points are deep enough, anywhere between about 90 kilometres (56 miles) and almost 1000 kilometres (600 miles), there is a chance that the electron will collide with an atom in the atmosphere. It is these electrons that cause the patchy pulsating aurora. It is common to see patchy aurorae in the early hours of the morning.

---

* *See also* diagram on page 180.

Of course, when we were skiing we never stayed out that late. Even on the night we saw the aurora, I was only outside for about twelve minutes before I retreated back into the (relative) warmth again, my fingers and toes already numb and sore. We also had to get up and ski during the day, so staying out all night, even if we could stand the cold, was not an option. It was hard enough to stay up to nearly midnight.

After the first couple of days of early nights and no checking for the aurora (it was cloudy anyway) we worked out a system. After we had eaten and defrosted some clothing, we would settle down into the sleeping bags for a rest. I would keep my face out in the cold and this would ensure I awoke later. Around eleven or eleven-thirty we would wake up again after a couple of hours' nap. We would chat, light the burner and heat up the water from our water bottles again. Then, bravely, we would dash out into the cold to go to the loo and check for the northern lights. This meant that we had a maximum of about a two-minute window, between the two of us, to see the northern lights. If they weren't out doing their dance during those two minutes then that was it for the night. We only saw them the once by this method. I realised that camping was not the optimal way to see the northern lights if I did not want to get out of the tent.

Temperatures steadily descended into the late -30°Cs. One morning when we were packing down the tent my toes were feeling numb and solid. Svante and I had been running around in small circles near the tent in an attempt to warm up and keep our feet from freezing. I put my face mask and goggles on. It was indescribably cold. We had to keep moving to stay warm. As soon as we had everything packed into the pulks we set off skiing, still wearing all our down clothing. After a while we would take off the down jackets, but we continued to ski in padded down trousers every day. Svante told me that in ten years of guiding he had never had to ski in padded trousers before.

I was not able to wear my goggles for long. They fogged up with millions of tiny ice crystals and clouded my view. So I took them off and endured my eyelashes freezing in exchange for being able to see the landscape. We still had not seen the Sun, though the sky took on a pleasing glow of orange. Whenever we stopped for a break the first thing I would do was remove a mitt and gently squeeze my eyelashes until all the blobs of ice had been melted.

In the evening when we got into the tent our clothing was stiff with frost. There was solid ice around the front collars of our waterproof jackets. We would remove the ice using a stiff brush, also removing our inner woollen mitts from their outer, windproof shells and brushing the frost off them. We had to remove as much ice as we could, then put gloves, hats, socks – everything we wanted to keep soft – into our sleeping bags, otherwise in the morning they would be frozen solid. Even wet wipes had to be sat on awhile before they could be used.

Thankfully we had the small burner (and extra fuel) to help keep us warm. It really did make a difference. At least it enabled us to take our mitts off without our fingers screaming in pain. In the light from our head torches, steam from boiling water and from our breath whirled in eddies. We spoke to each other through fine mist. This would freeze as rime on the inside of the tent so that every time we moved and disturbed it, ice crystals would fall on us as if it were snowing in the tent.

One night, quite late, I went outside to the loo and to check for the aurora. It was so cold that when I came back inside about a minute later I was shivering so violently that I started hyperventilating. I burrowed down into my sleeping bag to get warm and gain control of my breathing. That must have been the coldest night of the whole trip. As I lay curled up in my sleeping bag I thought how getting snowed-in in the Himalayas the previous year was like a picnic compared to this.

We reached the end of Reindalen and turned back. There had been talk of going all the way to the east coast and back, which I would have loved to do, but we didn't really have enough time for the round trip. We had the time a group would normally take to go one way, and though we were moving significantly faster than a group, averaging around 25 kilometres (15 miles) per day, my right knee had been playing up so we didn't want to push it. I thought about getting back to Longyearbyen in the next few days, how it would be fun to do some day ski tours without the pulks, or take a snowmobile ride somewhere. It would be nice not having to camp.

We began retracing our steps – or slides – back up Reindalen. That day I saw a beautiful cloud lined with orange and glowing from beneath from the Sun. It made me think about the Sun and why so many different cultures worship it. Being without the Sun for so long served to highlight a connection that is so strong yet so overlooked; I never realised how much I take the Sun for granted. Now I found myself longing for it, wanting to feel its rays on my skin, on my face, just to feel its warmth again. We continued on through the brightening twilight. It had been almost a week now since we had seen another person. Svante told me that no one goes skiing in January–February, or even into March. People thought we were crazy to do it. Most of the ski expeditions in Svalbard start in April under the midnight sun.

Maybe we were a bit crazy. I had wanted to see the northern lights in the wilderness, but I had no concept of what it would be like. I don't think you can, unless you have experienced it. If you have not camped out at almost -40° you have no benchmark.

'But I think you have to be out here to learn all these things,' Svante said. 'It is not enough to just talk about it; you have to experience it. Feel how cold it is and how sore your fingers will be, how the skin will crack up . . . all these things.' It truly was an incredible experience.

By the end of the week I was faster at tent duties. I had got into my routines and found the best ways of doing things, though I was still slow in the morning. During that beautiful but intense week I learned that the conditions the polar explorers had to endure were extreme. Out there in the wilderness, everything is about getting things done as quickly and efficiently as possible. If you stop you get cold. It's dangerous. You have to focus on doing just what is necessary. Out there, the simple, ordinary things that we fret about just didn't seem important anymore. All that mattered was the methodical performance of everyday tasks – changing socks, keeping warm, eating, drinking, skiing, smiling. It sounds dramatic, but simply living in those temperatures was enough. There was no space in my head for anything more frivolous. I was grateful to have Svante, with all his experience, looking out for me. It certainly gave me a greater appreciation of what explorers and scientists did in these frozen regions.

The last day was long and difficult. We skied from Reindalen back up the narrow river valley and over the pass where we camped the first night, then all the way back down Bolterdalen to the road. The last couple of hours we skied in the dark by head torch, keeping pushing for the road rather than setting up camp again. I was looking forward to not waking up with ice around the opening of my sleeping bag. The moon was up and the mountains were faintly illuminated.

We skied to a dog camp on the winding hill road up to Mine 7. It was hard work pulling up the hill after a long day. I was tired, my bad knee was hurting and my shoulders and back were stiff. I looked up at Svante ahead of me, dug my boots in and pulled.

Then he called to me, 'Look behind you!'

I turned and there it was, the northern lights. Faint but green, it stretched across the sky in a wispy arc, with multiple rays rising up to a zenith above us. I laughed and said a silent 'thank you' to the darkness. Then I turned and pulled. Within minutes I was at the top, skiing past a little house there, snowed-in and empty, past the frozen seal meat hanging high on a pyramidal rack out of reach of the bears. I could hear the howling and barking of dogs.

We unclipped our pulks and Svante called a taxi. We sat on the sleds and waited, watching, the aurora welcoming us back. I almost couldn't believe it. Then I saw the taxi lights coming up the road towards us, my guide's profile in silhouette. Within a quarter of an hour we were back in the brightness of town, in Svante's store room unpacking kit and hanging things up to dry. I felt whisked from one world to another. It was a strange feeling.

In Longyearbyen, when we had defrosted enough to feel again, I thought back over the previous week's skiing and even beyond, back across this entire journey. At the beginning, when I thought of the northern lights I imagined a beautiful story, one fed by films, fairytales and romance. But the Arctic is a different story. It is harsh and it is real, though no less beautiful. I thought back to the piled fairytale snow of Kiruna, the eagerness with which we climbed the ski slope to wait for the aurora. I'm still waiting for that spectacular display, for that perfect combination of conditions to set the sky alight. But perhaps the beautiful story is still possible – if I don't try to ski through a strange, stark land before the sun is up. Next time I'll be looking for a remote lodge with a log fire and a large window facing north.

* * *

THE DAYS PASSED, and if the first part of my trip was about survival, the second part was about regeneration. I watched the wind strip and tear, whistling audibly over the red plastic road markers, sending snow running across the roads in synchronised streams and swirls. The land took it all, icing up in its defence, glassy and hard. Then the wind would stop and the snow would fall, softening the landscape once more. Slowly I watched the Sun come back. I watched the days lengthen and weakly brighten.

In polar latitudes twilight is long. Even if the Sun is below the horizon it will not be completely dark, and all the snow reflecting the light means it often feels lighter than perhaps it would do without the snow. When I arrived in Svalbard in mid-February the Sun was still below the horizon and the island was in twilight, but there was more light than I expected. By the end of the ski trip a week later the Sun was peeking above the horizon, but not above the mountains so we never saw it. The most we saw was colour, diffused sunlight staining the skyline or tinting the snow. On 8th March the Sun was welcomed back to town with Longyearbyen's annual 'solfest'. Small crowds gathered at noon by the church in the old town, across the river from the main centre. Children sang and everyone looked hopefully south to a gap in the mountains where the sun peeps through and throws the season's first light on a small patch of ground in Longyearbyen. But it was cloudy. We saw the hazy rays trying to penetrate, but it was another two days before I saw shadows in town.

I did see the Sun once before that, however – on 4th March from the Kjell* Henriksen Observatory on Breinosa mountain. I was with a group of scientists from the Birkeland Centre for Space Science. We travelled together on a bus up the mountain, past Mine 7 to the EISCAT radars. From there we were taken up in two groups by a banded vehicle that could get through the snow to the observatory. We piled into the back, sitting along two bench seats on either side, wedging ourselves in by pushing feet hard up against the opposite side and holding on to a bar above when the ride became bumpy. There was a large window at the back. Suddenly as we approached the top of the mountain the cab brightened and someone said, 'there's the Sun.'

'Yes, the Sun!' I found myself laughing, 'I haven't seen the Sun in almost three weeks.' I leaned forward to look out of the back window to where the Sun was just visible over the opposite mountains. As soon as we stopped I leapt down into this new light. The sky was blue with wispy high cirrus clouds, the snow was sparkling and I could see my shadow – long, thin and pale, but definite. I faced the Sun and stood there squinting, trying to feel its heat on my face. It was too cold to feel its warmth, but I could feel its light and that was enough. I hadn't realised how much I had missed the Sun, even though I knew I wanted to see it. I was surprised at how excited I was that first time. If this is how I feel after three weeks, I thought, imagine how the people who live here must feel after six months of twilight and polar darkness.

---

* 'Kjell' is pronounced a little like a cross between 'shell' and 'chell'.

It is understandable that the people in Longyearbyen celebrate the Sun coming back to town. It is almost a primal feeling. The Sun is such an important part of life; it has featured in mythology and sometimes worship for millennia, also forming the stimulation for much of the early astronomy. The Vedic religion, the precursor to modern Hinduism, has one of the earliest written texts dating back to 1500 BCE. It contains hymns praising the Sun as the source and sustainer of all earthly life. Part of the Vedas was devoted to astronomy and was already an influential work by the fourth century BCE, possibly feeding into Greek astronomy. Many of the Christian Church's holy days are rooted in solar festivities – Christmas at the winter solstice, Easter at the vernal equinox. Richard Cohen, in his vast biography of our star and our relationship with it, *Chasing the Sun*, says, 'I have come to feel that the Sun has powers that are still well beyond our understanding – mythic indeed.' Nowhere is its influence more intensely felt than at the polar regions, both day and night, in light and in darkness.

The excitement of seeing the Sun dominated my visit to the Kjell Henriksen Observatory, much as I was interested to see its neat row of domes and the cameras inside. The KHO, as it is known, is the third dedicated auroral station at Longyearbyen, though auroral research on Svalbard goes back to the late nineteenth century.

The first auroral expedition to Svalbard was in 1882–3 during the first International Polar Year. Swedish scientists stayed in Isfjorden as part of a bigger research project across numerous locations to obtain measurements in metrology, geomagnetism and the aurora. They logged their observations, made measurements, recorded colours and attempted to measure altitudes, but these were highly variable. This was before Birkeland's expeditions and experiments, so knowledge of the aurora at this time was very limited. The team also performed some spectroscopy to analyse the colours of the aurora, but they didn't recognise any of the colours as coming from elements known on Earth. We now know that the elements involved are oxygen and nitrogen, and that the transitions involved – the particular drop from one electron orbit in the atom to another – only happen when the particle densities are much lower than those found on Earth. At the time they seemed like exotic new elements.

The next aurora expedition to visit Svalbard was to Akseløya in 1902–3, a small, narrow island at the mouth of a long fjord, Van Mijenfjorden, in Spitsbergen. This was one of the four sites of Birkeland's Aurora Polaris Expedition. Despite the first photograph of the aurora having been taken in 1892, aurora photography was still insufficiently developed for research purposes, so all observations were visual. As protection from the brutal weather, the scientist taking the measurements watched the northern lights from inside a large barrel. Nonetheless, they were thorough in recording shape and colour as well as estimated intensity, drift and speed.

It wasn't until 1978 that the first regular studies of aurora started on Svalbard. An international aurora expedition was organised on Spitsbergen and they built a small auroral station in Endalen, the valley alongside Longyearbyen. Svalbard is further north than the typical auroral zone derived from night-time viewing probabilities, so most aurorae are seen on the southern horizon and the archipelago is not regarded as one of the top places to see them. Yet for those who could stay through the winter, Svalbard offered a scientifically unparalleled view – the aurora during the daytime. The new station housed bespoke instrumentation capable of detecting the weak red colour of day-side aurorae.

Svalbard is a unique place to study the northern lights because its high-latitude position gives an interesting perspective. As we found out in Iceland, the Earth's dipolar magnetic field is not aligned exactly with the axis of the Earth, meaning that the magnetic poles are not aligned exactly with the geographic poles. The dipole is tilted towards Canada. At the magnetic polar regions the Earth's magnetic field drops to zero and creates funnel-shaped regions open to the solar wind and known as the polar cusps. Through these funnels, charged particles directly from the solar wind enter the Earth's atmosphere, creating aurora on the day side of Earth. These low-energy, unaccelerated particles penetrate less deeply into the atmosphere than those on the night side, resulting in a weaker light display that is predominantly red in colour.

Svalbard is located underneath the northern polar cusp and, due to the dipole tilt, occupies a more northerly position than other places under the cusp on the Canadian side. At such a high latitude, Svalbard experiences twenty-four-hour darkness in winter, so has the opportunity to see day-side aurora during polar night, when the sky is even dark at midday. Day-side aurora can only be seen for about two months around the winter solstice. The only other place in the world to see day-side aurora is Antarctica, which is much less accessible than Svalbard and where the communications are not as good. This is one of the reasons why Svalbard is so geophysically important. Although not the vibrant, dancing, night-time aurora that fascinates those of us from lower latitudes, these day-side auroral displays are particularly related to solar activity due to the direct access of solar wind plasma. Svalbard is a window into the solar wind.

In 1983 a more modern auroral station was built in Adventdalen about 6 kilometres (4 miles) from Longyearbyen, and it was expanded again only five years later. In Ny-Ålesund another research station opened so that simultaneous observations could be made from the two locations. In those days, Longyearbyen was simply a mining town run by Store Norske, whom researchers had to inform if they were visiting. Physicists were able to get a lift out to the old auroral station on the coal company bus with the miners. There were no hotels. Researchers stayed out at the auroral station, sometimes for months, though they could eat in town at either Huset or the miners' café, Busen. A permanent presence was required at the station throughout the aurora season and scientists had to be up all night to operate the instruments.

Huset, 'the House' – an imposing 1950s edifice built by Store Norske and standing alone and well lit on the edge of town – acted like the village hall in the mining days. There was a café, a small shop and a bridge room. Over the years Huset has been used as a school, a church, a hospital, a cinema, an airport terminal and in various other functions as need arose. Its size and grandeur, compared to the wooden cabin buildings making up most of the town, make it a local landmark. Now it houses a gourmet restaurant with an internationally recognised wine cellar, as well as a stylish but casual café-bar and a nightclub.

Kafé Busen was part of the bathhouse building in the centre of town, open 23 hours a day from 6am until 5am. Miners reporting for their shift would enter by one door, change their clothes and exit via another door, where a bus would be waiting to take them directly to the mines. At the end of the shift this happened in reverse – they would come in, shower and change before leaving via the 'clean' door. Longyearbyen has quite a culture of taking off shoes on entering buildings. At the university everyone walks around in socks or slippers, equally in guesthouses and even some restaurants and bars. I'm told this is a legacy of the mining days to keep out the dirt, rather than to keep out the snow.

The university, UNIS, opened in 1993, and the Radisson Hotel opened in 1995 (the building was brought out from the mainland after the 1994 Lillehammer Winter Olympics). Once there were hotels in Longyearbyen, the town grew. Scientists could now stay in town and make short visits to the auroral station for observing. No longer was it necessary for them to stay out there for weeks, alone with their cameras and instrumentation and the wide, dark skies.

As Longyearbyen developed, light pollution became a greater problem. A smeary, sodium-orange haze was seen on the pictures from the auroral all-sky cameras as light from street lamps reflected off low-altitude clouds. Research was also growing. It became clear that a solution was needed. In 2008 the new observatory opened. It was named the Kjell Henriksen Observatory after the leader of the original day-side aurora project in the 1970s.

'The largest observatory of its kind,' according to director Fred Sigernes, the KHO has thirty small units for cameras or other instrumentation. Around twenty different groups from ten countries around the world use the observatory, installing their equipment in a unit and operating mainly remotely from their home institutions, with perhaps a week's visit at the start of the observing season for maintenance and calibration. Their presence is no longer required at the observatory full-time; Svalbard has fast optical internet connection to mainland Norway so the data can be accessed, and the computers controlling the scientific instruments are able to turn on and off on their own at sunset and sunrise. There are various instruments operating, from year-round non-optical ones like magnetometers and radio receivers, to the optical instruments that only run in the dark winter season. As well as the all-sky cameras that give the fish-eye view of the whole sky, there are spectrometers to identify particular wavelengths of light; photometers that measure the intensity of individual colours across a full slice of the sky; and interferometers to look at one colour very specifically, seeing how its wavelength is Doppler shifted* and deriving information on the movement of neutral particles. All this to understand more about some wispy red lights in the sky.

Of course, we know there is more to it than that. Scientists don't just study the aurora because they are curious about the red light. We now know about the atomic transitions in oxygen that cause the red light, and all the other colours that we see in the aurora. We know how an atom or molecule in the atmosphere may be excited by an

* The Doppler shift is a change in wavelength of a wave caused by a moving source. It is the reason why we hear the pitch of an ambulance siren change as the vehicle passes. If the source is moving towards us the wave is squashed up, so the wavelength is shorter and the pitch (frequency) higher. If the source is moving away then the wavelength will be longer and the pitch lower. When measuring the Doppler shift of particles, or even galaxies, scientists measure how the wavelength of light emitted or reflected has changed from what it should be. Anything more red is moving away from us, more blue and it is moving towards us. This is what is meant by the term 'red shift'.

incoming particle enabling these transitions to occur. We even know something about where these incoming charged particles come from. We also know that we can't have night-time aurora without particles coming in on the day side. They feed the cycle. Particles scooped up on open field lines at the front of the magnetosphere are dumped into the tail region where they increase the density and squeeze the magnetic fields until the release of reconnection. Day side and night side work in harmony – we can't have one without the other.

More importantly, we know that, for the night side, true aurorae – what the world refers to when they speak of the aurora – where the particles come from originally is not the crux of the matter. Rather, it is the transfer of energy within the system, the acceleration of particles and the dynamics of the ionosphere that are the important elements of the aurora. The magnetosphere is receiving huge amounts of energy as it is buffeted by the solar wind and it cannot store it forever, it must do something to dissipate it somehow. What it does is deposit the extra energy in the atmosphere to make the aurora. To do this it accelerates particles within the magnetosphere and changes the travelling angles of their helical orbits so they can reach deep enough into our atmosphere to interact. This is a reproducible solution that happens in much the same way each time – albeit to different degrees. Just ask someone who has seen the aurora hundreds of times, like James Pugsley in Yellowknife, and they can describe to you the patterns of the substorm.

These important elements of the aurora are still not fully understood. Scientists know the general mechanism for how the aurora is produced, as described in this book, but there are recurring features that demonstrate that they still do not have the full picture – features such as hemispheric asymmetry, the overall morphology of the arc, or the mysterious transpolar arcs. In Longyearbyen I joined a meeting of scientists discussing some of these open questions. They were from the Birkeland Centre for Space Science, hosted at the University of Bergen and headed by Professor Nikolai Østgaard, but also including experts from around the world.

Most of the aurorae are broadly similar in both hemispheres, displaying the same patterns in the north and the south. These patterns can be studied from spacecraft that image the aurora in ultraviolet light, so that they can be seen even in daylight (a necessary requirement when one hemisphere is in summer and the other in winter). 'It's not uncommon for them to be twisted with respect to each other, but we do see the same features most of the time,' Karl M. Laundal from the University of Bergen told me. 'But then we have some very clear exceptions.'

The exceptions are thought to be to do with the variation in conductivity of the ionosphere, which changes as it is heated by the Sun's rays; or the fact that the magnetic field is not a perfect dipole, so field patterns can vary between the hemispheres and, of course, change over time. These are subtleties, but they tell us that we do not yet understand how the Earth, and the space around it, works.

Transpolar arcs tell a similar story of our ignorance. Also called 'polar cap aurora', 'weak Sun-aligned arcs' or 'theta aurora' because of the resemblance to the Greek letter, these aurorae occur within the dim region inside the auroral oval and stretch across the middle of this polar cap from day side to night side, almost bisecting it. They are rare and very faint, usually sub-visual even, although they were reported on early polar expeditions where detailed observations were made on a regular basis. The Swedish Expedition to Svalbard in the First International Polar Year produced clear drawings of the Sun-aligned arcs, and in the southern hemisphere the Australasian Antarctic Expedition headed by Sir Douglas Mawson also recorded the phenomenon. The unusual arcs can be seen clearly from space and again they remind us that we do not yet fully understand the evolution of the magnetosphere.

Aside from the magnetosphere, we don't yet fully understand our own atmosphere. We must remember that aurorae are part of a larger, global system. We may only see them in narrow rings around the polar regions, but that is the local effect of a global process. As such they may have a wider effect on our atmosphere, climate and weather as part of this system. Scientists are working to better understand how the magnetosphere interacts with the ionosphere and how these interactions propagate down into the lower atmosphere, where they can affect us here on Earth. The energy deposited in the atmosphere as the aurora will heat the ionosphere and change the proportions of neutral versus charged particles, which will modify the dynamics and the chemistry of the upper atmosphere. To quantify the effect requires interdisciplinary research between ionospheric physicists and atmospheric physicists. It is possible that the aurora could be affecting the weather, and if so, this is where the effect will originate.

'It's speculative,' said Hilde Nesse Tyssøy, from the University of Bergen, as she talked me through her research, 'but it could affect the surface temperature. If it does it will be a minor impact, it won't be a climate-changing thing, but it is a piece in the puzzle of how the natural solar variability affects us. I think most of the work on solar variability has moved away from the idea of global climate now and is thinking about the regional climate, and that's important. It matters for the weather and it matters for the ice.'

The research is still very new and the ideas controversial, but perhaps a scientific link will be found with the weather to support the impressions of Knut, the reindeer herder from Karasjok, who told me that they see more colours in the northern lights when the weather is changing.

In understanding this connection between the magnetosphere and the ionosphere, old-style rocket experiments like those launched from Kiruna before the space age are still immensely valuable. Scientists need to understand how the neutral particles behave and how they interact with the charged ions. Although the ionosphere is a plasma it is not fully ionised, meaning that neutral particles still remain in the mix with the free electrons and ions. Indeed, they form the majority. Radars such as EISCAT and SuperDARN can be used to track ions, but neutral particles are more difficult to measure because they don't respond to electric or magnetic fields or to radio waves. One way of measuring them is to release neutral coloured gas from a rocket into the aurora and see where it goes.

This past winter, a NASA mission was launched from Svalbard to do just that. NASA flew a small, twin-prop plane over to Svalbard on a long, multi-leg flight that took the pilots about a week to complete, so that the well-instrumented plane could photograph clouds of barium and strontium gas that were fired by rocket into the day-side aurora. Both gases are photo-luminescent, so they are visible in the sky – the barium a purplish pink and the strontium, blue. The barium ionises quickly while the strontium remains neutral, so the barium will align with the magnetic field and drift with the ions, whereas the neutral strontium will not. Using images from the aeroplane and from ground stations at Longyearbyen (the KHO) and Ny-Ålesund, the researchers can monitor the motion of the clouds using triangulation techniques, then ultimately derive the neutral and ion 'winds' to make a calculation of the interaction between different layers of the atmosphere.

Finally, a global system like the aurora and the magnetosphere needs a global view, and this requires a broad range of people with different skills. Also, speaking more literally, we could benefit from new global space-imaging missions so scientists really can see the aurora from new perspectives. If scientists are only looking at a small area, the theories proposed may be too narrow and may be a misrepresentation of the true situation. Jesper Gjerloev, a senior scientist at the Johns Hopkins University Applied Physics Laboratory, told me a story about a visit to Seville Cathedral in Spain.

Jesper has a theory about how currents flow in the ionosphere during auroral substorms. It is more complicated than the original picture, published by McPherron in 1973, but Jesper says that his is the global view. If you look at local measurements you can make a generalisation of what the world looks like, but when you have global measurements you may see that the whole picture is very different.

'I was in Seville a few weeks ago,' he said, 'and I was in the biggest Gothic cathedral in the world.'

He explained that he looked down at the floor and noticed he was standing on a beautiful star pattern. He took a picture of it. Then he looked up and realised that the entire floor is actually squares – a black-and-white pattern of squares and triangles.

'It blew my mind!' said Jesper.

To Jesper, the cathedral floor was a stunning visualisation of his model in relation to McPherron's. Only when viewed in isolation did small patches of the floor look like stars – the global view changed the picture.

With space-age technology we have the possibility of getting a global view. Scientists are devising new missions to make a huge leap in auroral imaging, such as the Ravens project proposed to the European Space Agency by an international team of scientists. It proposes two spacecraft with auroral cameras in polar orbits to enable continuous coverage of both hemispheres, something that is not possible with just one spacecraft. This 24/7 coverage will help scientists see the evolution of the magnetosphere. The mission is named Ravens after a legend in Norse mythology. The Norse god Odin had two ravens and he sent them out around the world to be his eyes and ears.

'We need a complete time history,' said Steve Milan of the University of Leicester. 'It is exciting to make prolonged observations in two hemispheres because there are open questions here.' Scientists expect the aurora to be similar north and south because the particles that are accelerated in the magnetosphere are bouncing on field lines between hemispheres. 'But the few simultaneous observations we do have say differently, and we don't know why,' concluded Steve, the lead of the Ravens proposal team, which also includes Eric from Calgary.

Study of the aurora, as with most other science, goes back to a fundamental desire to understand our surroundings, right out to the edges of the Universe. This kind of exploration, from the physical to the intellectual, is a fascinating characteristic of mankind. Through exploration comes knowledge, and through knowledge comes possibility. We don't know yet where this research will lead, but we do know that it will spawn further benefits for all of us – spin-offs from the insatiable desires of people to answer seemingly mundane questions, but which they will pursue with such fervour that they will develop new technologies to aid them in answering their question. It is in this way, stimulated by scientific questions, that imaging has developed to such an extent that we now all have small, powerful cameras in our phones, not to mention the capabilities of digital Single Lens Reflex cameras. We all rely on the Internet, which was developed initially to connect computer networks at research and university sites so scientists could share data more easily. Spin-offs from space exploration are almost too numerous to mention. NASA and collaborators have designed many new materials, with applications from car tyres to firefighter clothing, developed freeze-drying technology, invented the cordless mini vacuum cleaner and improved robotics that are being used for artificial limbs, to name just some. Fundamental research – simple exploration – occupies an essential place in the history and development of mankind.

Chatting in the canteen at UNIS one afternoon, I was treated to some stories of the old days at the observatory – of 'Stu with the flu' who visited one year and made everyone sick, of the Dane with the bike who would cycle out to the old aurora station in winter darkness through deep snow. They told me of how in bad weather the whole building would shake, wood creaking in the wind and the roof vibrating as if someone was walking along it.

'You can get a bit paranoid when you stay there alone for days and days and days,' I was told. 'There is no light, so you don't see anything. You look out the door and it's just pitch black and all the windows are covered because light will damage the instruments.'

Imaginations work overtime in these situations. During one storm, a student was so convinced that a polar bear was walking on the roof that he went out with a gun to investigate. But there was nothing. It was just the wind.

The stories that most amused me, however, were the anecdotes of national stereotypes. It started when they were telling me about how water at the Kjell Henriksen Observatory is a precious resource because of its position on the mountain. It is quite a mission just getting the water up there. First a snow plough has to clear the road up beyond Mine 7, which is usually travelled by a banded vehicle. The road clearance takes a few hours. Then water must be boiled in town and driven up by truck before it freezes. Waste water is even worse; that has to be driven down the mountain before it freezes.

Consequently, they have to be very careful of their water usage at the observatory. Not everyone realises this, especially the British, whose obsession with tea means that they use water like a free resource. The locals nicknamed us 'The Boilers'.

'Five Brits in the station boil more water than the other fifty people put together,' they laughed. 'Worse, they walk with the cup! There is constant movement with tea, and they drink it at their desks, by the instruments.' They sounded surprised, but it sounded completely normal to me.

'They could even fog up the domes in the old station,' one said. Fortunately there is much better ventilation in the new observatory. There was a less-than-helpful suggestion made by a Russian student: 'they should drink vodka!' Apparently vodka has the added advantage that it can be used to clean the windows.

'And the French! They are worse than the Brits,' the conversation continued. 'They bring their cheese and wine and make dinner while they are doing fieldwork. And not just one course, it's more like a seven-course meal!' One French group was famous at EISCAT because they would have fabulous lunches whilst at the radar station.

The Japanese are special, according to my companions. They don't eat or sleep, except for short naps sitting upright at their desks. 'They work more than thirty hours every day, even the students. And if you ask if they sleep they say "no".' It is, of course, fine to sleep. There are beds and sofas at the observatory for rest.

Then, of course, the Russians bring vodka. (Fred, the director, has quite a stash of it now.)

I was enjoying this insight into international life at the observatory. I laughed as I finished my tea.

* * *

PERFECT ALIGNMENT IN the heavens brought Longyearbyen a total solar eclipse on 20th March 2015. But though the celestial bodies may move in well-understood and predictable cycles, the behaviour of our troposphere is much more chaotic. Thus it was that for the few weeks preceding the eclipse, general discussions on the subject in Longyearbyen focused on hoping that it wouldn't be cloudy and worrying about how busy it would be in town with the flood of tourists. Longyearbyen is a town of around two thousand residents, with one supermarket and a handful of restaurants. The hotels had been booked up for years and there was even talk of the authorities opening up the sports hall to accommodate people who would come regardless. There had been a Norwegian airline strike while I was there, and before that a delivery boat hadn't been able to get into the fjord because of ice, so stocks of certain foods – milk, and fresh ingredients especially – were running low. Some of the bars in town had been serving a reduced menu for weeks. It was an inconvenience rather than a problem, but there was a lot of speculation about what it would be like over the few days around the eclipse.

A solar eclipse occurs when the Moon passes between the Sun and the Earth, thereby blocking the light of the Sun. The Moon orbits the Earth once every twenty-nine and a half days. This is the origin of the word *month* – a cycle of the Moon – and so the Moon passes between the Earth and the Sun once a month. At this point in the orbit the illuminated face of the Moon is pointed away from us, so we don't see it. We call it the *New Moon*. However, despite the Moon passing between the Sun and the Earth once a month, we don't see an eclipse every time. The Moon's orbit around Earth is tilted with respect to the Earth's orbit around the Sun by about five degrees, so they don't line up exactly – they cross. About twice a year the orbit planes will cross and an eclipse can happen somewhere on Earth. The window of opportunity is about 31 days, because the Sun's position in our sky shifts quite slowly relative to the Moon (our orbit takes a year; the Moon's takes a month), and the angular size of the Sun and the Moon in the sky means they don't have to line up perfectly for us to see something. The Sun and the Moon are both about the size of a pea if you were to try to grip them between your fingers. If the Sun is at the edges of the window of opportunity the Moon may just graze the side and produce a partial eclipse; towards the middle of the period the Sun can be blocked entirely. Where precisely they cross determines the type of eclipse we see.

Eclipses can either be total, partial or annular – that is, the disc of the Sun is fully obscured, partly obscured, or shows a bright ring around the edge of the Moon. An annular eclipse occurs because the two orbits involved are elliptical rather than circular, so the distances from Sun to Earth and Earth to Moon vary, slightly changing the perceived size of the Sun and the Moon in our sky over time. If the Moon appears smaller than the Sun in the sky then we will see an annular eclipse. This happens slightly more often than a total eclipse.

We are the only planet in the solar system to see a total solar eclipse, and this only happens by chance. All other planets with moons experience eclipses, but depending on the sizes and distances involved the moon shadows may simply be spots on the surface of the planet. Alternatively, they may block the Sun entirely, covering the corona as well. It just so happens that because of the size of our Moon and its distance to Earth relative to the Sun, when lined up the Sun and the Moon are about the same size. The Sun is four hundred times the diameter of the Moon, but also four hundred times further away from Earth. The Moon can obscure the Sun completely and we are able to see the beautiful solar corona. It is pure coincidence that allows us to see this spectacular view. And it won't last forever; the Moon is moving slowly away from Earth at a rate of about 4 centimetres (1.5 inches) each year.

During a total eclipse, where the Sun is fully obscured, the Sun, Moon and Earth line up *exactly*. In other words, a line drawn through the centre of the Sun and the centre of the Moon would hit Earth. Where it hits is called the *umbra*, the darkest part of the Moon's shadow, and it is only about 160 kilometres (100 miles) wide. Outside this region is the *penumbra*, a larger area of faint shadow where the Sun is only partly blocked out. As the Earth rotates, this shadow sweeps across its surface darkening a narrow (160-km/100-mile) track around 16,000 kilometres (10,000 miles) long. This path of totality covers less than one per cent of the Earth's surface. So although a total solar eclipse occurs about every one or two years, there is an average of 375 years between total eclipses happening in the same place. There have been erratic intervals in the past and there will be more in the future – for example, people living at the confluence of the Ohio and Mississippi Rivers in the US will see two total eclipses in seven years (2017 and 2024) – but in general a total eclipse is rare in your small corner of the Earth. The UK mainland will not see another total eclipse until 2090, the last occurrence being 1999 in Cornwall.

Like the northern lights, eclipses have been viewed since ancient times, often with a mixture of fear and awe. For most cultures, an eclipse boded ill. Folklore includes such stories as the Sun being eaten or destroyed, or the Sun god being angry and needing to be appeased. However, there are also some stories of love – the Sun and Moon lovers who come together periodically, their union in darkness, but who ultimately must part.

More than folklore, there is evidence to suggest that some ancient cultures understood the predictable cycles of eclipses, if not their exact cause. The most obvious of these is perhaps

Stonehenge in southern England. This circle of stone archways and independent marker stones was built and extended between 2800 and 1500 BCE. The alignment of the stones acts as an accurate solar calendar; its size, precision and laborious construction indicating its importance to the people who built it. 'These massive, shaped boulders were dragged from a quarry 32 kilometres (20 miles) away to codify in stone the discoveries of an earlier people,' says Mark Littmann et al in *Totality: Eclipses of the Sun*. Mayan and Chinese writings also indicate that these cultures knew about eclipse cycles.

Observations in more modern history go back to a description of the corona from 968 CE. Later, astronomers such as Kepler (in 1609) and Edmond Halley (in 1715) put forward their thoughts on prominences and the corona. The name 'corona' for the atmosphere of the Sun was popularised by Francis Baily in a description he wrote of the 1842 eclipse that he witnessed in northern Italy. Baily was 68 years old at the time and this was his third eclipse, but the previous two had been annular eclipses – when a ring of brilliant sunlight surrounds the Moon – and the corona had not yet been revealed to him in its true splendour. Baily had had a varied youth, exploring North America for two years after an apprenticeship in commerce, returning to become very wealthy from stockbroking, before retiring at 51 to devote himself to astronomy and in particular his fascination with eclipses. Baily is remembered in the field via the phenomenon known as Baily's beads – bright sparkles of light that occur around the rim of the passing Moon as sunlight streams through valleys between the mountains on the Moon's surface. On viewing the 1836 eclipse in Scotland, he had commented that the effect was 'a row of lucid points, like a string of bright beads.' The mid-nineteenth century was the beginning of expeditions being launched entirely to make scientific observations in whichever land in the world an eclipse was taking place, and Baily was one of the first eclipse chasers.

With the emergence of photography and spectroscopy, the pace of progress in understanding increased. The first photograph of an eclipse was taken in 1851. It was grainy and indistinct, but solar prominences and the inner corona were visible. By 1868 the spectroscope had been invented – scientists could break down light into a spectrum of lines and from this identify chemical elements, temperature and the density of the source. This opened up a new way of studying the celestial bodies far out of reach.

In the same year, 1868, both the British scientist Norman Lockyer and his French colleague Pierre (Jules) Janssen realised that by isolating individual spectral lines the Sun and its prominences could be studied at any time, not just during a solar eclipse. The idea to do this struck Lockyer two years earlier than Janssen, for whom inspiration struck while viewing prominences for the first time during the 1868 solar eclipse in India. However, Lockyer's experiment was delayed by construction of a new, better spectroscope for the task, so, coincidentally, the two reports of the similar discoveries were presented to the Academy of Sciences, in Paris, at the same session. Lockyer's studies of prominences led him to the discovery of helium, as mentioned in our discussion of the Sun in Iceland. Janssen's studies of eclipses led him unexpectedly into a new kind of astronomy. Two years later, during the Franco-Prussian war, Janssen escaped a Paris under siege by means of a balloon, just to view the 1870 eclipse in Algeria. This experience led him to make the first forays into high-altitude astronomy, studying meteor showers from balloons above the clouds. Both scientists continued to lead expeditions to see total solar eclipses for the remainder of the century.

In the twentieth century, a total solar eclipse was used to confirm Einstein's Theory of General Relativity. One of the predictions of this theory is that gravity will bend light in the same way that it will bend the paths of particles or other bodies made of matter. In the same way that a ball follows a parabolic path through the air when thrown, curving back on account of the gravity of the Earth, so light will also be minutely bent. The effect is very small, so immensely high gravity is needed. Only very massive objects will be capable of bending light rays enough for us to see this; a massive object like the Sun, Einstein realised. If we could block the light of the Sun to see the stars behind, their positions in the sky should appear slightly different to when viewed at night with the Sun out of the way. The light from stars just blocked by the Sun would be bent as if by a giant lens, and we would see them slightly to the outside of the Sun's disc, as if the light had been travelling straight. Back then, a total solar eclipse was the only time that the Sun's light would be blocked enough to see the stars. We now have coronagraphs to make artificial eclipses, and we can detect radio waves that incur the same bending, but then they were waiting for the right conditions during a solar eclipse. It was 1919 when the prediction was confirmed and Einstein became a celebrity. The effect he predicted is now called gravitational lensing.

In the early years, eclipses were studied to refine theories on the orbit of the Sun and the Moon; later to enhance our understanding of the corona and prominences; and in the beginning of the twentieth century to confirm an outlandish new theory. In modern times, despite our use of chronographs and space-based measurements to study the Sun continuously, observing natural eclipses from the ground is still worthwhile. Occulting discs that create an artificial eclipse also block out the lowest part of the corona, so we still see the full corona best during a total eclipse on Earth.

Consequently, scientists were visiting the UNIS group at Svalbard for the 2015 eclipse. They were predominantly from Hawaii, and they came to perform spectroscopy. There is still the huge mystery of why the solar corona is so hot, so scientists are diligently observing the Sun in the hope of piecing together the puzzle. Studying the Sun also helps us understand more about the basic laws of physics by allowing us to study conditions that would be impossible to create on Earth. For example, the number of particles per cubic metre in the corona is lower than even the very best vacuum we can achieve on Earth, so by studying the corona we can learn how matter behaves at such low densities.

There was general excitement at the university about the upcoming eclipse. The Department of Arctic Geophysics was getting ready to host scientific visitors, a general lecture course on the aurora and the eclipse was open to interested public as well as students, and researchers were finalising their own experiments. Fred Sigernes, Dag Lorentzen and Lisa Baddeley were preparing for an aeroplane flight which would give them a guaranteed view of the eclipse, even if there were clouds present. They would be testing a new hyperspectral imager and were hoping for a glimpse of the daytime, red aurora when the Sun was obscured. Seeing both the eclipse and the aurora would be a spectacular combination. Anticipation was high, but the weather had been mostly cloudy for several weeks now. I didn't really dare hope.

Eclipse day dawned clear. I couldn't quite believe our luck. I walked down through Longyearbyen towards UNIS just before seven o'clock. There was a large lenticular cloud hugging the summit of the mountain across the fjord, and some wispy, pink cirrus clouds up high. Otherwise the sky was blue.

I met my friend Pål* Ellingsen at UNIS and we drove across to the old aurora station in Adventdalen. Pål is an optics specialist in the Space Physics Group and he had a telescope and camera to set up. At the old aurora station, visiting scientists were setting up their equipment for imaging and spectroscopy, and the Norwegian broadcaster NRK was preparing for a live broadcast. The clouds disappeared and the sky became a perfect blue, the Sun shining strong and clear and reflecting brightly around the snowy mountain landscape.

Soon after nine o'clock a group of students arrived. Up and down Adventdalen people were gathering. It was cold – probably around -20°C – and I was wearing full down clothing. People walked around between small groups, trying to keep warm and anticipating 'first

* Pål is pronounced Paul.

contact'. Then, at about twelve minutes past ten local time, there it was. The Moon touched the Sun for the first time. Over the next hour we watched it move gradually across, people tracking the progress through their eclipse glasses. We also had a Sunspotter, a folded Keplerian telescope, which projected an image of the Sun onto a small screen. People gathered round to watch.

Gradually the Sun shrank to a small crescent, then a sliver. At this point it became noticeably darker.

'It's just like looking through sunglasses,' I heard a student nearby exclaim. Except we weren't wearing sunglasses.

Then, a strange strobe-like effect began, clearly visible as flickering light-and-dark on the snow. People were cheering and oooh-ing. These were the shadow bands – rippling waves of light caused as the final, almost point-source sliver of sunlight is focused and defocused as it passes through warmer and cooler air currents in the atmosphere. From what I had read about eclipses, I had been primed to watch out for the shadow bands because they are easy to miss. Experts recommend putting a white sheet out on the ground because they are most visible on a flat, white surface. On Spitsbergen, the shadow bands were obvious. We were surrounded by a flat, white landscape of snow.

Incredibly quickly, after seemingly only a few pulses of the strobe, there was a sudden brightening, known as the diamond ring effect, and then darkness. Totality. The Moon was completely blocking the Sun and I looked up at it with the naked eye. I could see the pink tinge of the chromosphere (so-called because of this distinctive pink colour) and prominences, though the fine structure of the prominences was more visible by telescope than by eye. I think I just saw the colour. The solar corona glowed silver, a ring around the dark shape of the Moon stretching short, silky fingers outwards into the black. It looked fairly symmetrical to me, not highly pinched in any direction. It was as if the Sun had taken on the Moon's sheen – an eerie, ethereal silver. I could see the odd star. It was beautiful.

I looked around at the mountains. With all the snow to reflect the light, the darkness was not a pitch black, more a navy blue. The mountains could be seen clearly and the horizon all around gave out a yellow glow. People stood staring up in awe.

All too soon there was another bright burst of light and strobing and the Sun was back. I fumbled in my mittens to open my eclipse glasses again and looked up to see the newly exposed crescent of the Sun as the Moon moved on its way. The Sun came back so quickly and sparkled through the landscape. The nearby star faded and the sky became light blue again. Even with the Sun as a mere sliver the world was once again bright. I took a deep breath as people around whooped and began chattering about how incredible it was.

I couldn't quite believe that I had witnessed a total solar eclipse. I felt elated. It had all worked perfectly. I was there in Svalbard, the weather cleared, the heavens aligned. To be in such a phenomenally beautiful setting in the wide valley, with the mountains around, the snow, the light – it was special.

After totality we watched the Sun return to us in full, everyone talking excitedly about what they had seen. Then at twelve past midday the Moon made last contact and was gone. Everyone packed up and left, leaving Adventdalen deserted once more. Apart from the scientists taking measurements from inside the old aurora station building, Pål and I were the last to leave, clearing up as you do at the end of a party.

* * *

THE WEEK BEFORE the eclipse, a lecturer at UNIS said that one cannot compare a partial to a total eclipse. Totality is unparalleled. It is true. I saw a partial eclipse near London in 1999 but I wasn't prepared for the difference I would see in Svalbard. What I found most interesting, most incredible, was the sudden transition from light to dark, from crescent to corona. It was the way in which the land went dark almost instantaneously that really struck me. One minute it was light, then strobing shadows, then a bright flash, then darkness. Then the Sun as you have never seen her. I have seen pictures of the corona before but, just like viewing the aurora, to witness it as part of the landscape gives a new depth to the experience.

Looking up at the corona and seeing the solar wind streaming away from the Sun, I was able to reflect on how intimately connected we are with the Sun on so many levels. I thought back over my journey and about all I had learned: from my first taste of the Arctic two years previously, to standing there in the snow on Spitsbergen seeing one of the most incredible sights on the planet.

My first view of the northern lights was in Kiruna, Sweden, when we learned that what matters for the aurora is the energy from the Sun and not the particles themselves. Then my journey took me to learn about folklore and history in northern Norway, the early auroral science and the beginning of our movement from myth to the possibility of knowledge. Iceland highlighted the geological structure of the Earth and how it furnishes us with a protective magnetic shield to keep out the worst of the Sun's radiation, without which life as we know it wouldn't exist. We learned about plasma and the Sun and its beautiful corona, and we found out about the streaming charged particles of the solar wind as we moved into Canada. We heard how images of the aurora can tell us about what is going on beyond our globe, about the intricate interplay of the solar wind and the Earth's magnetosphere, accelerating particles to Earth, and about how this familiar pattern is repeated time and again as a substorm, like the breathing in and breathing out of the magnetosphere. We learned about how light is created by the collisions of accelerated electrons with atoms in the upper atmosphere. In Scotland we realised there is a darker side to the northern lights. We found out about space weather and the adverse effects of large solar storms on technological systems and the human body, and how the world is beginning to work together to improve predictions and forecasting of space weather. Then here, to Svalbard, where we have learned about the structures of the aurora, and that there is still very much about the mechanism of the aurora that we do not understand.

That is the science, but there is much more than that, too. We have also learnt about the landscapes and cultures, the histories of people who devote their lives to the pursuit of an idea or a goal. We have learnt about the strength of the human spirit, our capacity for pain, endurance, loss and renewal. I've learnt about the fragility of these locations: hostile yet vulnerable, subject to change and with a future so uncertain. The world may change on the ground, but the lights will still be there above, going about their merry dance. By understanding all these different interactions we can better appreciate the beautiful phenomenon of the polar lights.

The aurora: faint and subtle, each one grows and strengthens, vague and changing and beautiful, sometimes imbued with our own feelings, imaginings and longings. But the aurora twists and writhes and ultimately disappears, leaving a ghostly imprint on the mind and a feeling of radiance and colour that we can't quite define or recall, but that we almost wish was still favouring us with her light. So it was. Now I go back to my world, carrying a big piece of the North with me.

Finally, I think I've come to understand better that feeling I had in Kiruna, when I felt that there was a peace about a snowy landscape that I didn't feel anywhere else. Now I know it's not about the snow as such; it's about the hostility. The soft, snowy beauty can change all too quickly to wind and ice and bare rock, to a place where, if you are not prepared, you may struggle to survive. These isolated towns are functional and utilitarian. Everything has a purpose. They still have the essential trappings of modern life – the Internet, transport, good food, central heating (my personal favourite) – but not the excesses, the paraphernalia, the superficial. Out here one may live well but simply. Out here it becomes starkly clear what is important in life. Therein lies the peace. Looking back through my history, I realise I've always chased this feeling, I've always been running. In the harsh, elemental North I now understand why.

> 'In the north are quivering arches of faint aurora, trembling now like awakening longings . . . never at peace, restless as the very soul of man.' Fridtjof Nansen

FURTHER READING

AKASOFU, SYUN-ICHI (2007). *Exploring the Secrets of the Aurora*. Springer Verlag, New York.

ALLISON, WADE (2009). *Radiation and Reason: The Impact of Science on a Culture of Fear*. Wade Allison Publishing, Oxford.

BREKKE, A. & A. EGELAND (1983). *The Northern Light: From Mythology to Space Research*. Springer Verlag, Berlin.

BREKKE, PÅL (2012). *Our Explosive Sun: A Visual Feast of Our Source of Light and Life*. Springer Verlag, New York.

BREKKE, PÅL & FREDRIK BROMS (2013). *The Northern Lights: A Guide*. Forlaget Press, Oslo.

CHOWN, MARCUS (1999). *The Magic Furnace: The Search for the Origins of Atoms*. Jonathan Cape, London.

COHEN, RICHARD (2011). *Chasing the Sun*. Simon and Schuster, London.

DWYER, MINDY (2007). *Northern Lights*. Sasquatch Books, Seattle.

EATHER, ROBERT H. (1980). *Majestic Lights:* The Aurora in Science, History and the Arts. American Geophysical Union, Washington, D.C.

GREEN, LUCIE (2016). *15 Million Degrees: A Journey to the Centre of the Sun*. Viking, London.

HUNTFORD, ROLAND (2009). *Two Planks and a Passion: The Dramatic History of Skiing*. Bloomsbury, London.

LITTMANN, MARK, FRED ESPENAK and KEN WILLCOX (3rd Edition) (2008). *Totality: Eclipses of the Sun*. Oxford University Press, Oxford.

LOPEZ, BARRY (2001). *Arctic Dreams: Imagination and Desire in a Northern Landscape*. Vintage, New York.

NANSEN, FRIDTJOF (1897). *Farthest North:* Being the Record of a Voyage of Exploration of the Ship 'Fram' 1893–1896. Archibald Constable and Company, London.

POPPE, BARBARA B. & KRISTEN P. JORDEN (2006). *Sentinels of the Sun: Forecasting Space Weather*. Big Earth Publishing, Colorado.

ROWLAND, HUGH (2010). *On Thin Ice: Breakdowns, Whiteouts, and Survival on the World's Deadliest Roads*. Hachette, London.

SUPLEE, CURT (2009). *The Plasma Universe*. Cambridge University Press, Cambridge.

INDEX

MW indicates Melanie Windridge (author).
Page references in *italics* indicate illustrations.

Abisko, Sweden 15–20
Abisko Scientific Research Station 15–17
ADEOS (Advance Earth Observing Satellite) II 214
Adventdalen, Svalbard 255, 273, 289–91
Akasofu, Syun-Ichi 152–3, 157
Alaska, US 2, 31, 88, 95, *96*, 124, 128, 152, 162, 212, 222, 250
Alberta, Canada 89–94, 96
Allen, Joe 230–1
Allison, Wade: *Radiation and Reason* 208, 209
Alouette I/Alouette II satellites 110–11
Alta, Norway *vii*, 25, 31–3, 34, 39, 53, 54
Ampère, André-Marie/Ampère's law 74, 75
Amundsen, Roald 30, 31, 51*n*, 56, 99
Angström, Anders Jonas 143
Antarctica 2, 13, 30, 86–7, 273
Appleton, Edward 175–6
Archer, Frederick Scott 107
Arctic Circle *vii*, 2, 46, 51, 51*n*, 128, 130, 166
Arctic Ocean 99, 128, 251, 254
Aristotle 43–4, 78
art, aurora appearance in 24, 192–3
Aston, Felicity 46, 86–7
Astronomy North 136, 138, 139
Athabasca, Alberta, Canada *vii*, 89–94, 97, 99, 123, 127, 128, 156
Athabasca University Geophysical Observatory 90–4
aurora (polar lights) 2; art, appearance in 24, 42, 192–3; auroral arcs 125, 149, 154, 191, 206, 262–3; auroral imaging and spectroscopy 52, 66, 85, 98, 117, 142–50, 271, 286–7, 288–9; auroral oval 13, 52, *53*, 94, 95, 98, 109, 110, 111–12, 113, 122, 125, 141, 152, 153–4, 156, 158, 222, 226, 228, 235, 262–3, 277; auroral science, history of 47–56, 62–86, 100–5, 109–10, 113–26, *120*, *123*, 142–59; auroral zone 1–2, 93, 94–6, *96*, 98, 99, 136–7, 162, 165, 166, 272; as a bad omen 44; Birkeland surmises causes of 52–4, 56, 109, 116–17, 271; calmness and wildness of 1, 3; classification/different forms 260–3, *261*; colours of 1, 2, 3, 21, 22, 37, 44, 143–51; composition of Earth's atmosphere and study of 151–2; day-side aurorae 18, 118, 121–2, 272, 273, 274, 279; early writings/literature and 24, 42–3, 45; electricity as cause of 48–9, 51; elusive nature of 2; favourable conditions for viewing 2; feelings of awe and spirituality in presence of 3, 24, 41–7; folk tales and legends, appearance in 24, 44–7, 48; formation of/basic science

behind 3, 17–20, 52–4, 62–86, 100–5, 109–10, 113–27, *120*, *123*, 142–59, 178, 180, *180*, 205–6, 231, 262, 264, 276, 280; height/location of 6–10, 49–51, 53, 109; light pollution and viewing 18, 20, 41, 90, 163, 164, 170–1, 274; most vivid occurrence, science behind 119–20; movement further south 3, 156–7, 158, 166, 192, 276, 277; MW fascination with 4–5; MW views for first time 4–5, 6, 10–22, 24, 292; mystique of 2–3; night-side aurorae 18, 121, 122, 152, 174; observing on other planets 150; photography and 7, 21–2, 50, 98, 105–12, 140, 141, 142–4, 149, 151, 154, 157, 163, 164, 165–7, 170, 178, 192, 205, 250, 260, 272, 279, 286–7; plasma and *see* plasma; polar explorers and 3, 24, 26–31, 46, 49, 51–3, 51*n*, 55, 56, 86, 108, 130, 133, 142, 251–2, 267, 268; prediction occurrence of/alerts for 248–9; reflections, interpreted as 44–5, 48, 51; Sami and 24, 34–40, 41, 45, 108, 134, 278; shapes and patterns of 3, 42, 50, 56, 100, 157, 260, 262, 272; solar wind connection discovered 117–18 *see also* solar wind; solar wind and magnetosphere, learning about conditions in from studying 150–1; southern lights (aurora australis) 2, 5, 156; space age advances science of 110–12; space weather and *see* space weather; spirits, interpreted as 37, 40, 45, 46; storms cause to move further southward 156–7; storms cause to occur over extended periods 157–8; substorms and 122, 123–5, 126, 136–7, 141, 151–7, 174, 222, 261, 276, 279, 292; sunspots connection discovered 48, 53 *see also* sunspots; term 4–5; winter-time nature of viewing 2–3, 18
aurora australis (southern hemisphere aurora) 2, 5, 156, 203
Aurora beer 198, 203–4
Auroral Imaging Group, University of Calgary 97, 111
AuroraMAX project 136–41, 151, 158, 159, 248

Baddeley, Lisa 289
Baily, Francis 286
Barentsz, William 251–2, 254
Bartels, Julian 234
Baschin, Otto 109
Beddington, John 183
Berg, Bredo 27
Berger, Tom 241, 242, 246
Bernoulli, Daniel: *Hydrodynamica* 70
Biblical references to aurora 42–3, 45
Birkeland Centre for Space Science 270, 276
Birkeland, Kristian Olaf 52–4, 97, 109, 110, 116–17, 142, 144, 178, 271, 272; *The Norwegian Aurora Polaris Expedition, 1902–1903* 53, 117, 181, 272
Björnsson, Gulli 88, 105
Bolterdalen, Svalbard 255–7, 268
Bossekop observatory, Norway 108–9
Botnar, Emstrur, Iceland 58, 68
Brendel, Martin 109
British Antarctic Survey 86
British Geological Survey 226, 243
Brunhes epoch 66
Brunhes-Matuyama reversal 66
Busch, Dr August 107

Caithness Astronomy Group 163–6, 170, 186, 195, 197, 249–50
Caithness, Scotland 161, 162, 164, 165, 166, 186, 189–90, 228, 249
Canada 2, 31, 88, 198, 200, 215, 262, 272, 273, 292; aurora links with throughout history 96–7, 162; aurora research in 91–4, 97–100, 110–11, 123–4, 125–6, 135–42, 156, 158, 159, 248–9; auroral zone and 94–6, *96*; Dene people and 40–1; geomagnetic storms and 216, 217, 218; Inuit in 45
Canadian Space Agency 136, 137, 151, 159
Cape Canaveral, Florida 177, 212
Carrington, Richard/Carrington event, 1859 103–4, 169, 172, 174, 183–4
Carter, Jenny 247
Castlehill Heritage Centre, Castledown, Scotland 163–4, 170–1, 249, 250
Cecilie 25, 31, 32, 34, 35, 36, 38, 39, 54, 134

Celsius scale 82*n*
Celsius, Anders 49
Central Radio Propagation Laboratory, Boulder, Colorado 177, 202, 232
Chapman-Ferraro Cavity 115
Chapman, Sidney 114–15, 117, 181
Christiania (now Oslo) University 52–4, 109, 144, 178
Chukchee people 45
Church, Frederic Edwin: *Aurora Borealis* 192
climate change 5, 15–16, 37, 133
clouds 7–8, *10*, 136, 166
Cluster missions, 1996 and 2000 5
Cohen, Richard: *Chasing the Sun* 271
collective behaviour 72–3, 77
Collodian process 107
Con Mine, The, Northwest Territories 129
Connors, Martin 90, 91, 92, 93, 129
coronal mass ejections (CMEs) 102–4, 119, 155, 167, 169, 170*n*, 173, 174, 185, 212, 213, 217, 218, 222, 230, 231–2, 233, 234, 237, 239–41, 243–4, 247
cosmic radiation 209
Cox, Allan 66, 67
CUTLASS (Co-operative UK Twin Auroral Sounding System) 86

Daguerre, Louis-Jacques-Mandé/ daguerreotype 106, 107
Dalrymple, G. Brent 66
Deep Space Climate Observatory, The (DSCOVR) 239, 240, *240*
Dene people 40–1, 128, 131, 132
Denison, John 130
Doell, Richard 66
Donovan, Professor Eric 19, 96, 97, 98–9, 110, 111, 112, 125–6, 136, 139, 151, 155, 156, 157, 261, 263, 280
Doppler shift 275, 275*n*
Dungey Cycle 121–3, *123*, 151, 203, 222
Dynamics Explorer 1 satellite 111–12

Earth: atmosphere 3, 5, 6–10, *10*, 13, 14, 17, 19, 37, 42, 50, 56, 68, 69, 76, 79*n*, 83, 84, 86, 88, 100, 102, 104, 110, 112, 113, 114, 115, 117–18, 121, 142, 143, 144, 146, 147, 148, 149, 150, 152, 157, 159, 168, 172–3, 176, 180, *180*, 185, 201, 202–3, 209, 211, 247, 264, 272–3, 275–6, 277–8, 279, 292; birth of 60–1; core, composition of 63–4; magnetic field/magnetism *vii*, 2, 18, 19, 51*n*, 52, 53, 54, 56, 58, 61, 62–77, *65*, 86, 87–8, 89–126, *96*, *114*, 120*n*, *120*, 142, 145, 146, 147, 150, 151–2, 153, 154–6, 158–9, 173–4, 178–81, *180*, 183, 184, 185, 192, 201, 202–3, 204–6, 208, 209, 210, 211–12, 213, 217, 218, 220, 222, 223, 224, 226, 231–5, 237, 238, 239, 243, 244–5, 248, 262, 263, 264, 271, 272–3, 275, 276, 277, 278, 279, 280, 292; plasma and *see* plasma; plate tectonics 61, 67; solar wind and 113–16, *114*, 118, 119–23, 154, 173, 181, 222, 234–5, 292; space weather and *see* space weather
earthquakes 48, 61, 63, 82
Eather, Robert H 124: *Majestic Lights* 43, 44, 109
Eddy, John 168, 169
Edmonton, Canada 93–4, 127
Egeland, Alv 109–10
Eiffel Tower 175, 176
Einstein's Theory of Special Relativity 75, 287–8
EISCAT (European Incoherent Scatter) radar 13–14, 98, 270, 278, 282
electricity grids, space weather and 217–29, 243
electromagnetism: development of theory 52–4; plasma and 72–7; space weather and 184, 201, 206, 208
Ellingsen, Pål 289, 291
Engholm, Sven 34–5
European Space Agency (ESA) 14, 247, 280
exosphere 9, *10*
Explorer satellites 111–12, 177–8, 180–1
explorers, polar 3, 24, 26–31, 46, 49, 51–2, 52*n*, 55, 56, 86, 108, 131, 142, 251–2, 267, 268
Eyjafjallajökull volcanic eruption, Iceland, 2010 58, 67–8, 182–3, 211

Faraday, Michael 73–5, 122
Ferraro, V.C.A. 114–15, 181
Finland 25, 134, 149
Finnmark, Norway 25, 34–5, 53, 134
Finnmarksløpet 25, 31–3

folk tales and legends, aurora appearance in 45–7
Fox Talbot, William Henry 106, 107
Fram (ship) 24, 27–9, 30, 192
Franz Josef Land 29, 108
fusion 4, 77, 78, 79, 80–2, 83, 116, 202

Galilei, Galileo 5, 47, 78
geodynamo 66
geomagnetic field, Earth's 52, 63–7, 90, 94, 99, 104, 113, 115, 120*n*, 205–6, 212, 222, 264
geomagnetic storms 103–4, 116, 117, 141, 147, 154–9, 172–85, 202–47, 248, 292–3 *see also* space weather
Gjerloev, Jesper 279–80
Glatzmaier, Gary 66
GPS (Global Positioning System) 110, 110*n*, 172, 204, 205–6, 214
Great Trigonometric Survey, The 49–50, 205
Greenland 16, 27, 28, 30, 45, 96, 251, 252

Haggarty, Ewan 245
Haldde mountains, Norway 53, 54
Halley, Edmond 48, 49, 286
Hammerfest, Norway 39, 172
Hapgood, Mike 182–3, 184
Heaviside, Oliver 175
heliosphere 100, 103, 235
Hippocrates 43
Horrebow, Christian 78–9
Hrafntinnusker, Iceland 59, 60
Hudson Bay, Canada 45, 95, 128
Huntford, Roland: *Two Planks and a Passion (The dramatic history of skiing)* 31
Hydro-Québec 217–18, 221, 224

Ice Age 26, 42, 62, 251
Iceland *vii*, 46, 53, 96, 105, 146, 292; CUTLASS radar and 86–8; Earth's magnetism and 62–8, 222; Eyjafjallajökull eruption, 2010 and 58, 67–8, 182–3, 211; formation of landmass 60–1; geological activity in 57–8, 59–62, 67–8, 292; Laugavegur trail hike 58–62, 88; plate tectonics and 61–2, 67–8; prime spot for viewing aurora 87–8; rhyolite rocks 60; Sagas 162–3

Inter-Service Ionosphere Bureau, British 176
International Auroral Atlas 95, 260
International Geophysical Year, 1957–8 152–3
International Polar Year, 1882–3 108, 271–2, 277
International Satellites for Ionospheric Studies 111
International Space Station (ISS) 7, 9, 10, *10*, 159, 200–1
Interplanetary Magnetic Field 100, 114
Inuit 26, 31, 41, 45, 51*n*, 131, 132
ionisation 9, 71–2, 143, 146, 206
ionosphere 9–10, *10*, 14, 86, 93, 98, 110–11, 113, 122, 173, 175–7, 185, 201, 202, 203, 204, 205–6, 207, 222, 232, 233, 235–8, 276, 277, 278–80
ISIS I and ISIS II satellites 111
ITER machine 77

Janssen, Pierre (Jules) 287
Jeannette 28
Jennings, Kerensa 46
Johansen, Hjalmar 29
Johns Hopkins University Applied Physics Laboratory 279
Joint European Torus 77

K-index 234–5, 243, 250
Karasjok, Norway 34–5, 38, 134, 278
Kármán line 6, 10
Kautokeino, Norway 108–9
Kennelly, Arthur 175
Kepler, Johannes 47, 286, 290
Kew Observatory, London 103, 174
Kiruna Geophysical Laboratory 13 *see also* Swedish Institute of Space Physics
Kiruna, Sweden *vii*, 4, 6, 7, 11–15, 20–2, 139–40, 269, 278, 292, 293
Kjell Henriksen Observatory (KHO) 270, 270*n*, 271, 274–5, 279, 282
Knudsen, David 263
Knut (Sami reindeer herder) 34, 35–40, 45, 134, 278
Kodiak Launch Complex, Alaska 212

Lagrange points 233, 239–42, *240*, 245, 249
Landmannalaugar, Iceland 58, 60, 62

Langmuir, Irving 72
Lapporten (the Sami Gate), Abisko National Park 15
Larmor, Joseph 178
Last Mooseskin Boat, The (documentary) 4
Laugavegur trail hike, Iceland 58–9, 67, 88
Laundal, Karl M. 277
Lee, Chris 149
Legislative Assembly, Yellowknife 129, 131–5
Leirvogur Observatory, University of Iceland 88
light pollution 18, 20, 41, 90, 163, 164, 170–1, 274
Lindemann, Frederick Alexander 117
Lockyer, Norman 85, 143, 287
Longyearbyen, Svalbard *vii*, 251, 253, 254, 255, 267, 269, 270, 271, 272, 273–5, 276, 279, 283, 289
Lorentzen, Dag 289
Lossiemouth, Scotland 197, 200, 204, 229
Love, Jeffrey 192

MacDonald, John: *The Arctic Sky* 41
Mackenzie River 89, 128, 130, 131
Mackie, Gordon 161, 163–6, 167, 170, 171, 195, 249, 250
Maddox, Richard Leigh 107–8
magnetic field lines, aurora and 3, 17, 18, 48, 53, 66, 75, 100, 101, 104, 116, 118–
23, *120*, *123*, 124, 125–6, 142, 150–2, 155, 156, 178, *180*, 205–6, 231, 262, 264, 276, 280
magnetic fields/magnetism *see under individual planet name*
magnetic poles *vii*, 2, 31, 51*n*, 67, 94, 99, 145, 272
magnetic reconnection 118–24, *123*, 146, 151–2, 157, 173–4
magnetopause 114, *114*, 115, 180–1
magnetosheath 114, *114*
magnetosphere 18, 19, 88, 93, 112–26, *114*, 146, 147, 150, 152, 155, 156, 157, 159, 173, 174, 180, *180*, 181, 202–3, 209, 210, 211, 212, 213, 222, 231–2, 234–5, 239, 244, 245, 264, 276, 277, 278, 279, 280, 292
magnetotail 124, 180, 262, 263
Mairan, Jean Jacques d'Ortous de 48, 49, 192
Mandan Native Americans 45
Marconi, Guglielmo 175, 176
Mariner II 181
Maunder Minimum 47, 166, 168
Maxwell-Boltzmann distribution *70n*, 71
Maxwell, James Clerk 70–1, 75, 107
McGarvie, Dave 62
McLennan, John 144
mesosphere 8, 9, *10*
Met Office 185, 204, 243; Space Weather Operations Centre, Exeter, UK (MOSWOC) 185, 204, 232, 242, 242*n*, 243
Milan, Steve 112–13, 126, 280
Milnes, Arthur 229, 230
Mitchell, Mike 40, 132–3
molecular behaviour 69–71, 70*n*, 148–9

Nansen, Fridtjof 24, 26, 27, 28–30, 46, 56, 192, 294
NASA 28, 78, 83, 110–12, 124, 126, 169, 170, 179–81, 209, 211–12, 223, 232–3, 232*n*, 233, 239, 241–2, 279, 281
National Grid, UK 185, 217–28, 242, 243–5
National Oceanic and Atmospheric Administration (NOAA) 181, 230, 232, 233, 233*n*, 235, 239, 243
Niépce, Joseph Nicéphore 105, 106
noctilucent clouds *10*, 136, 166
Norberg, Carol 11, 12, 13
Nordland, Odd 46
Norse 44–5, 162–3, 186, 187, 252, 280
North Magnetic Pole 31, 51*n*, 94, 99, *120*, 120, 120*n*
North Pole *vii*, 27, 28, 29, 120, 207
Northeast Passage 108
northern lights (aurora borealis) *see* aurora
Northwest Passage 51–2, *51*, 99, 130, 133
Northwest Territories, Canada 40–1, 127–34; Northwest Territories Council 131–2
Norway 23–56, 108, 109, 162, 171, 172, 253, 275, 289, 292; aurora and folklore of 44–6, 292; Birkeland and 53–4; dog sledding and skiing in 25–34; explorers and 26–31; MW

winter training in Oslo 54–6; Sami in 34–40, 45, 136 *see also* Svalbard *and under individual place name*
Ny-Ålesund, Svalbard 254, 273, 279

Olmsted, Denison 52
Omholt, Anders 109–10
Onsager, Terry 237–8
Østgaard, Professor Nikolai 276
OVATION Aurora Forecast model 235
ozone layer 8, *10*

Parker, Gene 117, 119, 181
photography, auroral 7, 21–2, 50, 98, 105–12, 140, 141, 142–4, 149, 151, 154, 157, 163, 164, 165–7, 170, 178, 192, 205, 250, 260, 272, 279, 286–7
Pioneer 3 and 4 missions 178, 179
pitch angle scattering 180, *180*, 264
plasma/plasma physics 4, 5, 97, 100–1, 112, 113–23, *120*, *123*, 146, 152, 155–6, 168, 169–70, 173, 174, 181, 185, 201–2, 219, 222, 292; behaviour/nature of 69–77; Carrington event and 169–70; CMEs and 185, 222, 230–2, 233, 234, 237; corona and 86; electron acceleration and 263, 264; fast and slow 222, 235, 237; 'frozen in' condition 100–1; geomagnetic storms/space weather forecasting and 155, 168, 169–70, 173, 174, 181, 222, 230–2, 233–5, 237–41, 244, 245, 246, 248; ionosphere and 201, 278; magnetic reconnection and 119–23, *120*, *123*, 126, 152; magnetosphere and 112–16, 118–19, 234–5, 264; plasma frequency 201–2; radiowaves and 93, 201–2; solar magnetism and 84; solar wind and 100–1, 102, 104, 114–15, 118, 168, 233–5, 273; Sun as 58, 78, 79, 80, 83, 84, 86, 93, 100–1, 102, 103, 104; sunspots and 83
polar cap aurora 277
polar cusps 117–18, 203, 210, 272–3
Poppe, Barbara: *Sentinels of the Sun: Forecasting Space Weather* 205
Prince of Wales Northern Heritage Centre, Yellowknife, Canada 40, 132
Pugsley, James 135–42, 151, 153, 154–5, 158–9, 165, 248–9, 276

quantum physics 145–6, 149, 150
Québec, Canada 95, 217–18, 221, 224, 243

radiation 8, 9, 114, 125, 173, 178–9, *180*, 181, 185, 204, 208–14, 230, 233, 264, 292
radio communication/radio waves 40, 93, 173, 175–7, 185, 201–6, 207, 208, 216, 230, 231, 232, 233, 234, 275, 279, 288
radioactivity 66, 116, 208, 230
radiometric dating 66
Read, Al 197–201, 203–4, 214–15, 216
Reindalen, Svalbard 257–60, 266–8
reindeer 27, 34, 35, 36–40, 45, 46, 134, 252, 256, 257, 278
revontulet (fox-fire legend) 44, 56
Reykjavík, Iceland vii, 58, 58*n*, 62*n*, 87, 88
Richards, Andrew 217, 218, 225, 227, 243
Royal Air Force (RAF) 197, 198–9, 204
Russia 25, 26, 45, 51, 53, 107, 108, 110, 134, 177, 253, 254, 282, 283
Rutherford Appleton Laboratory, Oxfordshire 5, 182, 211
Rutherford model of atom 145, 145*n*

Sami people 15, 24, 34–40, 41, 45, 108, 134, 278
satellite communications services, space weather and 204–7 *see also under individual satellite name*
Scandinavia 2, 6, 47, 88, 96, 98, *124*, 134, *153*, 162, 163, 187 *see also under individual nation name*
Schuster, Arthur 116–17
Schwabe, Heinrich 79
Scotland *vii*, 2, 94; forecasting space weather and 197–201, 225–50; history of human habitation in 186–9; observing northern lights in 163–7, 170–2; Scandinavian history of 162–3; standing stones and stone circles in 187–8
Scott, Captain Robert Falcon 30–1, 46, 47, 56
seismology 63, 82
Seville Cathedral 279–80
Shrum, Gordon 144
Siberia 2, 28, 45–6, 88, 96
Sigernes, Fred 274–5, 289

Sinclair, Chris 170, 171–2, 186, 187–8, 195, 250
skiing 25, 26–7, 28, 29–31, 55–6
sledding 25–6, 28, 29, 30, 31–3, 41, 129, 255–60, 264–9
solar eclipse 84–5, 101, 103, 107, 283–92
solar energetic particles (SEPs) 212–13
solar maximum 88, 102, 104, 105, 152, 156, 159, 167, 168–70, 176, 184
solar storms 103–4, 116, 117, 141, 147, 154–9, 172–85, 202–47, 248, 292–3 *see also* substorms *and* space weather
solar wind 19, 83, 86, 88, 100–5, 112–23, 114, *114*, *123*, 125, 150–1, 154, 155, 156–7, 159, 168, 173, 181, 222, 232, 233, 234–5, 237, 247, 272–3, 276, 292
South Magnetic Pole 86, 120*n*
South Pole *vii*, 30, 46, 55, 87, 120
space weather 172–3, 292–3; astronauts and 211–12, 213; Carrington event, 1859 103–4, 169, 172, 174, 183–4; cosmic radiation and 208–12; damage to satellites and other equipment 211–14, 245, 246–7; enhanced aurorae, storms and 155–9; flight crew and 209–11, 213; flights over poles and 207–8, 210; forecasting solar storms 230–42; geomagnetic storms 103–4, 115, 141, 147, 154–9, 172–85, 202–48, 292–3; GPS and 204, 205–6, 214; Halloween storm, 2003 214, 223–4; history of observance 174–6, 183, 216; knowing how to utilise forecasts 242–7; military operations and 204–5; NASA launches and 211–12; power transmission systems/ electricity grids and 216–28; radiation storms 173, 185, 204, 208–14, 233; radio communication and prediction of 176–7; radio communications interference and 201–3, 204; satellite communications services and 206–14; satellites and forecasting 245, 246–7; solar flares and 173, 185, 212–13, 217, 230, 231; space age and 177–82; spacecraft use in forecasting 245–6; term 181; UK government's Risk Register and growing awareness of 172, 182–6, 216; voltage swings, stabilising 243–4
Space Weather Prediction Center, (SWPC), Boulder, Colorado 177, 202, 232–4, 235, 237, 239, 241, 242, 243, 246, 248; International Space Environment Service (ISES) 237
Spaceport Sweden 14
spectroscopy 52, 66, 85, 98, 117, 142–50, 271, 286–7, 288–9
Spitsbergen (Svalbard) 53, 54–5, 252, 253, 272, 290, 292 *see also* Svalbard
STEREO satellites 170, 233, 241–2
Stonehenge 285–6
Store Norske Spitsbergen Kulkompani 253–4, 273–4
Størmer, Carl 109–10, 178–9, 192, 205, 260
Strand, Svante 55, 255–9, 265, 267–9
stratosphere 8, *10*, 172
substorms 122, 123–5, 136–7, 141, 151–6, 157, 174, 222, 261, 276, 279, 292
Sun 48, 53, 56, 77; atmosphere 77, 79*n*, 83–6, 100–5, 117, 143, 286, 288–90 *see also* corona *and* solar wind; chromosphere 84; corona 84–6, 143, 286, 288–9; coronal mass ejections (CMEs) 102–4, 119, 155, 167, 169, 170*n*, 173, 174, 185, 212, 213, 217, 218, 222, 230, 231–2, 233, 234, 237, 239–41, 243–4, 247; fusion and 80–2, 83–4; Interplanetary Magnetic Field and 100; layers of 79–80, 82–3; light emissions 82–3; magnetic field 83–6, 100–5, 167; photosphere 82–3, 84; plasma and 69, 77, 84; size of 78, 79*n*; solar eclipse 84–5, 101, 103, 107, 283–92; solar energetic particles (SEPs) 212–13; solar flares 103–4, 167, 169–70, 173, 185, 212–13, 217, 230, 231; solar magnetism 84–5, 100–5; solar maximum 88, 102, 104, 105, 152, 156, 159, 167, 168–70, 176, 184; solar storms 103–4, 116, 117, 141, 147, 154–9, 172–85, 202–47, 248, 292–3 *see also* substorms *and* space weather; solar wind 19, 83, 86, 88, 100–5, 112–23, 114, *114*, *123*, 125,

150–1, 154, 155, 156–7, 159, 168, 173, 181, 222, 232, 233, 234–5, 237, 247, 272–3, 276, 292; sunspots *see* sunspots; temperatures 81, 82–3, 82*n*, 84, 85
sunspots 47, 48, 53, 78–84, 102–5, 116, 156–7, 167–70, 174, 230–1, 241, 244
SuperDARN radars 86, 278–9
Svalbard, Norway 53, 54–5, 108, *124*, 172, 230, 251–92; cultural heritage 254–5; discovery of 251–2, 254; first auroral expedition in, 1882–3 271–2; first photograph of aurora taken in, 1892 272; hunting and trapping in 252–3; KHO in 274–5; mining in 253–4, 255; MW sees northern lights in 258–61, 264–5, 268; MW ski trip in 255–61, 264–9; NASA aurora mission in 279; regular studies of aurora in 272–83; second auroral expedition in, 1902–3 272; Sun's appearance in 269–71; Svalbard Treaty, 1920 253; total solar eclipse in, 2015 283, 288–92; unique place to study northern lights 272–4, 279; whaling and 252, 252*n*
Sweden: MW views northern lights for first time in 4, 6, 7, 11–22, 24, 292
Swedish Institute of Space Physics 13, 14
Sweet, Peter 119

Takasaka, Yuichi 166–7
terrella 53–4, 109, 116, 142, 178
THEMIS mission 123–5, 126, 136, 138, 141, 156
theodolite 49–50
thermosphere 8–9, *10*, 173
Thomson, J. J. 116, 145
Thurso, Scotland *vii*, 162, 163, 166, 195, 197
Tiddy, Nigel 197–9
Torfajökull, Iceland 60, 62
triangulation 49–50, 53, 98, 109, 205, 279
Tromholt, Sophus 108–9, 110; *Under the Rays of the Aurora Borealis* 109
troposphere 7–8, 10, *10*, 172, 283
Tyssøy, Hilde Nesse 278

United States Geological Survey 66, 192
University Centre, Svalbard (UNIS), Norway 254, 274, 281, 288–9, 291
University of Bergen 276, 277, 278
University of Calgary 19, 96, 97, 99, 110, 111, 125–6, 136, 151, 263
University of Leicester 112, 247, 280
Urdahl, Laurentius 26–7, 30, 31

Van Allen, James/Van Allen radiation belts 177, 178, 179, *180*, 181, 264
Vegard, Lars 144
Voyager 1 100

Walport, Sir Mark 242
Watt, David and Kitty 189–95, 228, 249–50
Weyprecht, Carl 108
Windswept Brewing Co. 197–9, 203–4, 214–15
Winiarczyk, Maciej 166, 171, 250
Wolf, Rudolf 79
World Meteorological Organisation (WMO) 238
WSA-Enlil 235, 248, 249

Yellowknife, Canada *vii*, 40, 127–42, 151, 154, 158–9, 165, 248, 276

ACKNOWLEDGEMENTS

Throughout this journey I have been truly humbled by the people I have met: people who gave me their time, their knowledge and their friendship so freely. Of course, my gratitude extends beyond, to those who influenced me and inspired me when this book was not a book, just a seed of interest in my mind. Yet what follows is a list of those who aided me in the immediate journey, from Sweden to Svalbard and from concept to manuscript.

Kerensa Jennings for sharing her experiences of the northern lights and providing inspiration and imagery for the opening of the book.

Carol Norberg, organiser of the Kiruna Arctic Science course, and friends from that first experience.

Cecilie Andersen and family for making me so welcome in northern Norway.

Engholm Husky for putting me in touch with Knut the reindeer herder, and to Knut for sharing tales of the aurora, his family and his livelihood.

Four English guys at the bus-stop in Reykjavík – Chris, Kit, Alex and Jonathan – who allowed me to invade their party and with whom I walked the Laugavegur Trail, enduring (and enjoying!)

long days of heavy bags, endless lava fields, volcanic dust and river-wading. Their company was appreciated.

Felicity Aston for showing me the sights near Reykjavík - the Golden Circle and the CUTLASS radar!

Gunnlaugur (Gulli) Björnsson in Reykjavík, for tea and cake and physics chats.

Dave McGarvie, Open University in Edinburgh, for information on Iceland geology and volcanology.

Eric Donovan at the University of Calgary, for extensive discussions on Canadian auroral studies, history, plasma physics and more besides, and for facilitating visits to Athabasca and Yellowknife. Thank you also for the copy of Bob Eather's book *Majestic Lights,* which proved a useful reference and good jumping-off point for further study.

David Knudsen for information on auroral morphology.

Martin Connors for accompanying me to Athabasca and showing me the observatory and the cameras.

James Pugsley for fun, interesting and informative evening chats in Yellowknife.

Steve Milan, University of Leicester, UK, for discussions on the magnetosphere.

Robert Fear, University of Southampton, for discussions on the science of the magnetosphere and reconnection.

Gordon Mackie and the Caithness Astronomy Group for coming to share stories at short notice: Chris Sinclair, Stewart Watt, Maciej Winiarczyk, John Hilton, Alan Kennedy, Karen Dunn, Harry Harvey, Lucy Munro. Additional thanks to Chris Sinclair for taking me out on an archaeological tour.

Kitty and David Watt for welcoming me into their home and sharing their art and their off-grid lifestyle.

Al Read at Windswept Brewing Co. in Lossiemouth for teaching me a bit about beer as well as talking about his Royal Air Force background and aurora experiences.

Arthur Milnes for stories of the RAF and encounters with the aurora.

Chris Towne for Colorado, and the writing top.

Dave Pitchford for numerous contacts in Boulder.

Terry Onsager, Tom Berger and colleagues at the Space Weather Prediction Centre, Boulder, USA – Bill Denig, Robert Steenburg, Howard Singer, Rodney Viereck, Bob Rutledge among others – for discussions on space weather and forecasting.

Janet Green and Jennifer Gannon for information on space weather and natural hazards.

Jeffrey Love for a brief history of geoscience over breakfast (and a ride to the airport) and a preview of a fascinating article on art and the aurora.

Joe Allen for email discussions on aurora on the declining side of the sunspot cycle.

Mark Gibbs and Catherine Burnett at the Met Office Space Weather Operations Centre, and others at the L5 mission conference, particularly Andrew Richards from the National Grid, Ewan Haggarty from the Airbus DS SKYNET programme, Markos Trichas of Airbus and Chris Lee from the UK Space Agency.

Mike Hapgood for information on the growing concern of space weather within government.

Jenny Carter, University of Leicester, for space stories, discussions, encouragement and unfailing support, always.

Svante Strand and Helen Turton, of expedition company Newland AS, for more than I can say. It was sometimes tough, but it was special.

Nikolai Østgaard (University of Bergen) and Lisa Baddeley (The University Centre in Svalbard) of the Birkeland Centre for Space Science meeting in Longyearbyen, who welcomed me in to listen to their talks and group meetings, and to join their social activities. Special mentions to those who provided extra information through further discussions: Jesper Gjerloev, Kjellmar Oksavik, Karl Laundal, Hilde Nesse Tyssøy, Dag Lorentzen and Anja Strømmer.

Rebekka Nilsson for companionship in a cold, dark Longyearbyen.

Pål Ellingsen and Ivar Marthinusen for friendship, skiing and some spectacular photographs.

My agent, Diane Banks, and Robyn Drury for nurturing the project and improving my work through thoughtful edits.

Myles Archibald, Julia Koppitz and the rest of the team at HarperCollins for work on edits, images, covers, titles and everything else that turns the manuscript, finally, into a book.

Jim Martin for giving that little seed the warmth it needed to germinate, setting me off on this crazy journey.

Atul Srivastava for reading and feedback.

Imperial College, particularly Steve Rose and Roland Smith (Plasma Group), for their support over the years, and Steve Swartz (Space and Atmospheric Group) for allowing me to gatecrash some space plasma meetings and lectures when the subject was of interest to me.

David Kingham and everyone at Tokamak Energy for enduring my lengthy disappearances.

And to my family, for their love and support even when the places seem wild and remote and the projects equally so.

Thank you to everyone here for believing in me enough to give me your time and energy. I am truly indebted.

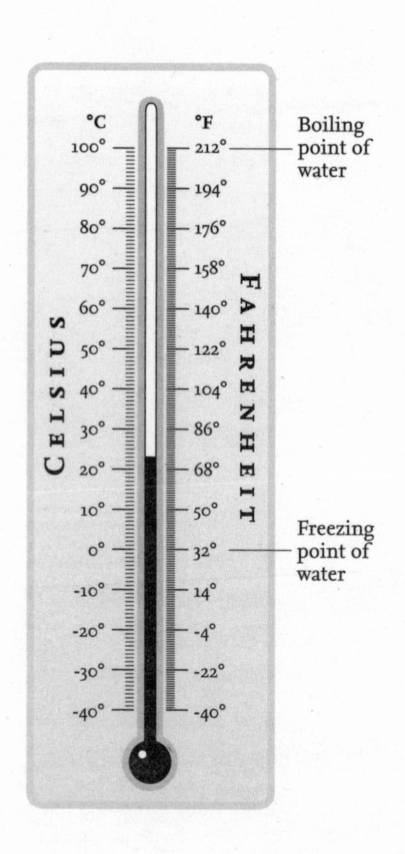
°C
°F
Boiling point of water
Freezing point of water
CELSIUS
FAHRENHEIT
100°
90°
80°
70°
60°
50°
40°
30°
20°
10°
0°
-10°
-20°
-30°
-40°
212°
194°
176°
158°
140°
122°
104°
86°
68°
50°
32°
14°
-4°
-22°
-40°